JN169091

やさしい 機械英語

Easy English for Young Mechanical Engineers

青柳忠克・齋藤哲治・塚原隆裕［共著］

はしがき

現在，日本の工業技術は世界で最高の水準にあります．世界中から多くの技術情報が日本へ入ってくると同時に，日本の技術情報が世界の隅々に流れていきます．それらの情報の交換の手段として英語が最も多く使われていることは，皆様すでにご承知のとおりです．

機械技術は，電気工学，建築工学，化学工学など多くの学問の基礎として極めて重要なものです．したがって，私達の身の囲りには機械技術を応用した製品，たとえば自動車，ジェット機，家電製品などがたくさん目につきます．

皆様が機械技術に関する情報を英文で読み，また研究成果を英文で発表する機会は将来ますます増えるものと思われます．その際にお手伝いができればと思って，本書「やさしい機械英語」を書きました．

この本の特長は次のとおりです．

技術の学習と英語の学習とが同時並行的に行えるように配慮しました．これまでの参考書の多くは，英語と技術はそれぞれ別の本に記載してありました．そのようなむだを排除するために，本書では英文法を説明すると同時に，機械工学の知識を基礎から管理に関する内容まで幅広く網羅しました．

第Ⅰ章の「機械工学の基礎」では，中学生程度の理科および数学の内容をやさしい英文で記述しました．専門用語さえわかれば中学生でも読解できます．第Ⅱ章・第Ⅲ章は機械工学本来の内容を高等学校程度の英文で記述してあります．第Ⅳ章・第Ⅴ章では最先端の機械技術および管理技術を取り上げました．英文はやや難しく，高専・短大・大学でも耐えるようになっています．

何事もそうですが，ことに英語は一朝一夕には成りません．息の長い勉強を続けてください．必す成果があがります．ご成功をおいのりいたします．

平成 28 年 10 月

著者らしるす

改訂２版の特長について

本書はこれまで，大学および高等専門学校において科学英語や技術英語の基礎的な教科書として活用されてきました．このたびの改訂にあたって，教科書としての使い勝手を向上させるために，新たに以下の特長をもたせています．

(1)　各項目がスッキリと見やすくなるように編集すると同時に各項目の順序を少し見直し，授業の進行状況に応じて利用しやすいように編集した．また，新しい項目として燃料電池，太陽電池，ハイブリッド車などの項目を追加した．

(2)　各項目の最後に内容に準じた問題を，さらには各章末にまとめの章末問題を配置した．復習用のテストとして利用できるように，問題および章末問題の解答はすべて巻末に記載した．

(3)　本文中の文法については，第Ⅵ章として新たな章にまとめ，基本的な文法内容について詳しく解説した．

(4)　英語に関するコラムを随所に配置し，語法や文法などの知識の充実が図かれるよう工夫した．

以下に，各項目の内容を示します．

●イントロ●	各項目の導入または説明．読む前の参考に．
語　句	重要な語句と難しい語句の発音と意味を記載．
構文	複雑な英文や難解と思われる英文の解説．
解説	機械の専門に関する内容の解説．
問　題	各項目の内容に関する練習問題．
章末問題	各章のまとめとしての英文理解と英作文の練習問題．
章末問題解答	章末問題の解答．
練習問題解答	練習問題の解答．
課題英文要約	第Ⅲ章以降の各項目の要約．
索　引	主な専門用語および文法用語を日本語と対照して記載．
文法（第Ⅵ章）	英文法の説明．必要に応じて参照のこと．

目　次

Ⅲ　機械工作　Machining

Ⅳ　機械工学の現在　Advanced Mechanical Engineering

I

機械工学の基礎

Basis of Mechanical Engineering

機械工学という膨大な学問の分野は，物理学，化学，数学などの多くの基礎的な学問によって支えられている．

この章ではそれらの基礎をできるだけやさしい英文で，番号をつけて解説してある．解説欄の同じ番号を対照しながら学習するとわかりやすい．

1 エネルギー

Energy

1. Many machines, such as motors, lathes, and automobiles work with energy.
2. Energy exists in a variety of forms.
3. Energy can be used as heat, electricity, mechanical energy, chemical energy, radiation, and so on.
4. In a battery, chemical energy is transformed into electrical energy.
5. Motors change electrical energy into mechanical energy.
6. Moving air, i.e. wind, or a moving car possesses mechanical energy.
7. Electrical energy results from the motion of tiny particles called electrons.
8. Many sources of energy on the earth originate from the sun.
9. Millions of years ago, plants and animals began to capture energy from the sun and store it in their body.
10. Coal and petroleum provide energy which was originally radiated from the sun a very long time ago.

語句

lathe[leið] 名旋盤　exist[igzíst] 動存在する　**variety**[vəráiəti] 名多様性，変化　**electricity**[ilektrísəti] 名電気　**radiation**[reidiéiʃən] 名（光・熱などの）放射　**transform**[trænsfɔ́:m] 動変形させる，変化させる　**i.e.**[ai–i:] すなわち，換言すれば（id est の略で that is と読んでもよい）　**result**[rizʌ́lt] 名結果　動…から起こる（from），…に終わる（in）　**particle**[pá:tikl] 名少量，微粒子（elementary～素粒子）　**source**[sɔ:s] 名源泉，出典　**capture**[kǽptʃə] 動捕獲（する）　**store**[stɔ:] 動貯蔵（する）　**provide**[prəváid] 動供給する　**originally**[ərídʒənəli] 副本来は

●イントロ●　エネルギーは現代の生活に欠かせないものとなっており，エネルギー資源の奪い合いからしばしば戦争が起こる．エネルギー資源をどこに求めるかは我々人類にとって永遠の課題であろう．

▶3. be used ➡「be 動詞＋動詞の過去分詞」の形で，受動態（受け身）といわれる．「(エネルギーは) 利用される」．

▶3. and so on ➡「…など」．and so forth と同じ（etc. と略される）．

▶4. …is transformed into～ ➡「be 動詞＋動詞の過去分詞」の形．「…は～へと変えられる」．

▶7. results from～ ➡「電気エネルギーは電子と呼ばれる微細な粒子が移動することにより発生する」の意．

▶9. began to capture…and store it ➡この to は capture と store との両方の動詞にかかっていることに注意．「動植物は太陽からのエネルギーを捕捉し，そして貯蔵することを始めた」．

▶10. energy which was originally radiated from the sun ➡関係代名詞 which で energy を詳しく説明している．「…はもともと太陽から放射されたエネルギーを供給している」．

解説 石炭，石油，天然ガスなどを化石燃料（fossil fuel）という．これらは，大昔の動植物が蓄えた太陽エネルギーが化石の形で保存され，現在利用されているものである．化石燃料は値段が安く，利用しやすいエネルギーであるが，燃焼に伴って大量の二酸化炭素や二酸化硫黄などが空中に放出され，さまざまな形で公害の原因となっている．将来はよりクリーンなエネルギー源への変換が求められるであろう．

問　題

1. 次の（　）の中の語を並べ変えて正しい英文とし，それを日本語に訳せ．
   (1) Energy (heat, be, as, and, electricity, can, used).
   (2) Chemical energy (electrical, is, energy, transformed, into).
   (3) Electrical energy (from, the motion, results, electrons, of)
2. 本文の内容と合っているものには○，合わないものには×をつけよ．
   (1) Automobiles and lathes work with energy.
   (2) Moving air possesses chemical energy.
   (3) Motors change chemical energy into mechanical energy.
   (4) Many sources of energy originate from the earth.

2 自由落体の速度

Velocity of Free Falling Bodies

1. Take a feather and a coin, and drop them from the same height.
2. The coin may fall faster than the feather.
3. This test may give you the impression that heavy bodies fall faster than light ones.
4. This view was the basis of a theory postulated by the ancient Greek philosopher, Aristotle.
5. This theory has been accepted to be true nearly for 2000 years.
6. In the late sixteenth century, the great Italian scientist, Galileo Galilei, questioned this theory.
7. He actually carried out experiments to test whether the theory was true or not.
8. Galilei's painstaking experiments disproved Aristotle's ideas, and led to build a solid foundation for modern science.
9. Since then, the experimental method has become very important for the study of natural science.

語　句

velocity[vilɔ́siti] 名(運動などの)速さ　**feather**[féðə] 名(鳥の)羽毛　**impression**[impréʃən] 名印象，感想← impress 印象づける　**view**[vjuː] 名視力，観念，計画　**postulate**[pɔ́stjuleit] 動仮定する　**philosopher**[filɔ́səfə] 名哲学者　**scientist**[sáiəntist] 名自然科学者← science サイエンス　**experiment**[ikspérimənt] 名実験　**whether**[hwéðə] 接…かどうか(多くは後に or not をつける)　**painstaking**[péinztèiking] 形苦心した　**disprove**[disprúːv] 動(…が)誤りであることを証明する　**foundation**[faundéiʃən] 名基礎，ファンデーション

●イントロ●　速度に velocity という難しい語を使っているがこれは speed（速度）とほぼ同義である．現在の科学ではガリレイが示した実証的態度が重視される．

構 文　▶**1. Take a feather…, and drop them**➡このように主語なしで，動詞を先頭に出す形を命令文という．内容を簡潔に表現できるので使用法や実験法の記述によく用いられる．「…を用意しなさい．それらを同じ高さから落としなさい」の意．

▶**2. …may fall faster than〜**➡形容詞の比較級の用法．「…は〜よりも早く落ちるかもしれない」．

▶**3. impression that heavy bodies fall…**➡この that は接続詞の用法で，that 以下で impression の内容を説明している．「重い物体は軽いものよりも速く落ちるという印象」．

▶**4. Greek philosopher, Aristotle**➡ Greek philosopher（ギリシャの哲学者）と Aristotle（アリストテレス）とは同格の用法．同じ内容を別のことばで表現している．「ギリシャの哲学者であるアリストテレス」．

▶**7. whether…or not.**➡このように whether は or not と対にして用いることが多い．この or not はなくてもよい．「…であるかどうか」．

▶**9. has become…**➡現在完了の継続の用法．「…の方法は非常に重要となり現在でも（引きつづいて）重要である」の意．

解 説　**アリストテレス（BC 384〜322）**は，ギリシャの偉大な哲学者であり自然科学者でもあったが，彼の学説には間違っているものも少なくなかった．ここに述べられている「重い物体は軽いものよりも速く落ちる」というのもその一つである．彼の学説は中世にあってキリスト教と結びついて非常な権威をもっており，「アリストテレスによれば……」というのが当時の知識の最大のよりどころであった．BC は紀元前の意．**ガリレオ（1564〜1642）**は，アリストテレスの学説に疑問をもち，有名なピサの斜塔（Fig. 1）を利用して自由落体の実験を行ったと伝えられている．

ガリレオはまた，天体観察の結果から地動説を唱えた．そのために宗教裁判にかけられ，地動説の放棄を命ぜられると同時に，自宅に軟禁され研究活動を禁止された．彼は裁判で判決を受けたと

Fig. 1　ピサの斜塔

き，「それでもなお地球は動いている」とつぶやいたといわれる．実験と実証を重んじる彼の主張が今日のあらゆる科学の基礎になっているといっても過言ではない．

問　題

1. 次の（　）の中の語を並べ変えて正しい英文とし，それを日本語に訳せ．
   （1）　Heavy bodies (light,　faster,　fall,　than,　ones).
   （2）　He actually (experiments,　carried,　the theory,　out,　to test).
   （3）　The experimental method (very,　become,　important,　has).
2. 本文の内容と合っているものには○，合わないものには×をつけよ．
   （1）　Aristotle was a famous Italian scientist.
   （2）　Light bodies fall faster than heavy ones.
   （3）　Experiments are very important methods for studying science.
   （4）　Galileo Galilei is called the father of modern science.

コラム　同格（Apposition）

同格とは，同じ事物を別のことばで説明する用法で，カンマ（,）や of で結んだり，また直接結ぶこともある．例えば「rake angle, α, …」とあれば，rake angle と α とは同一のものを指し，互いに同格という．同格関係としての補足説明の部分はカンマで挟まれることが多い．カンマの後に，namely や or または that is（to say）を付けて「すなわち」や「言い換えると」のニュアンスを伝えることもある．名詞と名詞を of で繋げて同格を示す方法もある．「the city of Tokyo（東京という都市）」

3 摩擦

Friction

1. When it is very cold, you may rub the palms of your hands together to generate warmth.
2. In ancient times, people rubbed wooden sticks together to start fires.
3. In Japan, people used the "Hinoki" tree, which means "fire tree".
4. Sparks are often produced when a tool is sharpened on a grinding wheel.
5. These phenomena are all caused by friction.
6. Friction wears out machines and slows their speed.
7. Sometimes friction can be useful to prevent the wheels of automobiles or airplanes from slipping.
8. However, sliding surface should have as little friction as possible.
9. To reduce friction, lubricant is usually applied.

語 句

rub[rʌb] 動こする　**palm**[pa:m] 名手のひら　**spark**[spa:k] 名火花　**grinding wheel** グラインダー（研削盤）の円盤　**phenomena** [finámənə] 名 phenomenon（現象）の複数形　**friction**[frikʃən] 名摩擦　**slow**[slou] 動遅くする　**prevent** [privént] 動妨げる，防止する　**reduce**[ridjú:s] 動減らす，減少させる　**lubricant**[lú:brikənt] 名潤滑剤

●イントロ●　摩擦は運動する物体の間に発生し，その運動を妨げるものである．摩擦をなるべく少なくしたい場合と，逆にそれを利用する場合とがある．

▶**1. When it is…，4. when a tool is…** ➡これらの when は「…するとき」の意で，接続詞の働き．「非常に寒いとき」，「工具を研ぐとき」．

▶**3. the "Hinoki" tree, which means "fire tree".** ➡ which は関係代名詞の継続用法．「そしてそのことばは"火の木"を意味する」．

▶**7. to prevent the wheels…from slipping.** ➡ prevent と from とは関連し

ている．「車輪を滑ることから防ぐために」，つまり「車輪が滑らないようにするために」．

▶**8. as little…as possible.** ➡二つの as は関連して用いられている．「できるだけ摩擦を少なくする」．

問　題

1. 次の（　）の中の語を並べ変えて正しい英文とし，それを日本語に訳せ．
   (1) Rubbing (the palms, generates, of your hands, warmth, together).
   (2) Friction (the wheels, slipping, prevents, from, of automobiles).
   (3) Lubricant (to reduce, is, friction, applied, usually).
2. 本文の内容と合っているものには○，合わないものには×をつけよ．
   (1) People rubbed wooden sticks together to start fires.
   (2) "Hinoki" means "fire tree" in Japanese.
   (3) Friction can be useful in some cases.
   (4) Lubricant can increase friction.

コラム a/an の使い分け

不定冠詞の a と an の使い分けは，直後に続く単語の（綴りではなく）発音に依存する．つまり，通常は a で，母音発音の前に限り an となる．

an ultraviolet ray	a unit（an unit ではない）
an LED light （エルイーディー）	a LAN cable （ラン）
an EU society （イーユー）	a European ally （ユーロピアン）
an n-th degree equation	an (x, y) plane

4 曲ったてこの周りのモーメント

Moment around a Bent Lever

1. Ordinarily, a lever is a straight rod, and sometimes a bent lever is useful in a machine.
2. In using a claw hammer to pull out a nail, as shown in Fig. 2, it is thought to work as a bent lever.
3. In this case, the principle of moments applies.
4. Let us suppose that you have to exert a 200 N (≒20 kgf) force on the hammer in Fig. 2 to draw the nail out of the board.
5. We can calculate the power x exerted on the nail.

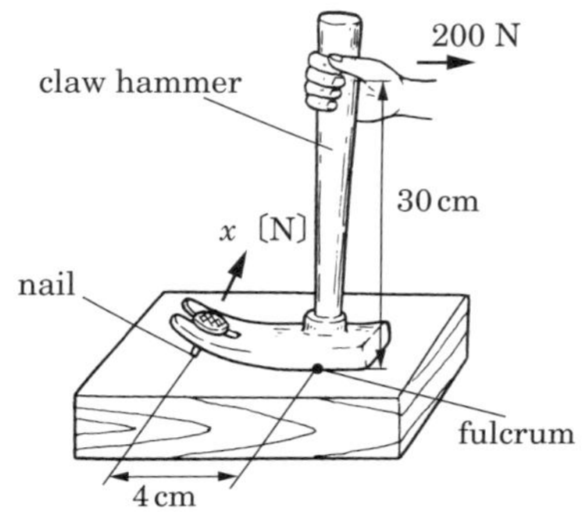

Fig. 2　Claw Hammer

6. The length of the handle from your hand down to the fulcrum is 30 cm.
7. The distance from the fulcrum to the nail is 4 cm.
8. Applying the principle of moments, we can calculate

   $30 \text{ cm} \times 200 \text{ N} = 4 \text{ cm} \times x \qquad x = (30 \times 200)/4 = 1500 \text{ N}$
9. We can obtain a mechanical advantage in force as large as 1500 N even though we apply only 200 N with our hand.

語　句

moment[móumənt] 名モーメント，能率　**straight**[streit] 形まっすぐな　**bent**[bent] 形曲がった←bend 曲げる　**claw hammer** 名釘抜きハンマ　**nail**[neil] 名釘，爪　**principle**[prínsəpl] 名原理，主義　**exert**[igzə́:t] 動用いる，働かせる　**board**[bɔ:d] 名板，テーブル　**calculate**[kǽlkjuleit] 動計算する　**length**[leŋθ] 名長さ　**handle**[hǽndl] 名柄，ハンドル　**fulcrum**[fʌ́lkrəm] 名（てこの）支点　**distance** [dístəns] 名距離　**advantage**[ədvɑ́:ntidʒ] 名利点，有利　**even**[í:vən] 形平らな　副…さえ，…でも

●イントロ●　てこ（lever）の原理は，はさみや栓ぬきなど，日常に応用される以外に機械にもいろいろな所で利用されている．わずかな力で大きな力を得ることができる．

▶2. In using a claw hammer ▶8. Applying the principle of moments ➡このように～ingを用いた構文を分詞構文という．「釘を抜くのに釘抜きハンマを用いるとすれば」，「モーメントの原理を応用すると」．前者のinはなくてもほとんど同意であるが，あればわかりやすい文になる．

▶2. as shown in Fig. 2 ➡「図2に示されているように」．

▶4. Let us suppose that…➡これは一種の命令形で自分自身に対する命令．「（that以下のことを）考えてみよう」．

▶6. from your hand down to the fulcrum ▶7. The distance from the fulcrum to the nail ➡ fromとtoとは関連していて，「…から～まで」の意．

▶9. as large as 1500 N ➡ asとasとの間に形容詞を入れたこの形は強調のため．「1500 Nもの大きい力」．

▶9. even though…➡ even ifとほぼ同意．「たとえたった200 Nの力をかけたときでさえも」．

解説　ペンチやプライヤを利用すれば素手では曲げられないような針金でも容易に曲げることができる．これはてこの応用であって，手の何十倍もの力を針金に作用させることができる．ある有名な学者は，「私に大きなてこと支点とを与えてくれれば地球を動かしてみせる」といっている．

問題

1. 次の（　）の中の語を並べ変えて正しい英文とし，それを日本語に訳せ.
   （1）　A bent lever (useful,　in,　a machine,　is).
   （2）　The point (a fulcrum,　is,　holding,　a lever,　called).
   （3）　We (on the nail,　calculate,　the power,　can,　exerted).
2. 本文の内容と合っているものには○，合わないものには×をつけよ.
   （1）　A lever must always be a straight rod.
   （2）　We gain an advantage in energy with a lever.
   （3）　A claw hammer means a hammer which can be used to pull out nails.
   （4）　We can pull out a nail using a claw hammer.

コラム　will と be going to の違い

未来のことを表す際に，will または be going to が用いられるが，そのニュアンスはまったく異なる．前者は助動詞であり，書き手・話し手の主観的な気持ちが加わっている．名詞 will には意思・精神力・意欲などの意味をもつことからも，will は主観的で積極的な印象をもった未来の表現方法である．これに対して，後者は通常，現在形の形をとり，淡々と客観的事実を言い表す方法である．

　I will go to university next year, and so I must study very hard !
　（来年は大学へ行くつもりだから，一生懸命に勉強しなければ！）
　I'm going to university next year, and so I've reserved a student dorm.
　（来年は大学へ行く予定のため，学生寮を予約した.）

他にも「（義務・約束・任務などにより）～することになっている」を表すために，be supposed to も用いられる.

5 液体中の圧力

Pressure in Liquid

1. You all have seen a car suddenly stopped.
2. In such cases, the driver must have stepped on his brake pedal.
3. Let's look at what happens when he steps on the brake pedal.
4. At first, the liquid stored in the brake system, the brake oil, receives great pressure, which is transmitted to the wheel.
5. The movement of the car body is stopped by an immense pressure exerted by the liquid under pressure.
6. When you pump up a flat tire, you inject high pressure air into the flat tire.
7. In an automobile repair shop, cars are easily raised by means of a lift operated by oil under pressure.
8. In all these devices, the oil or air under pressure is conveniently used to transmit pressure to a remote place.
9. When pressure in liquid is studied, Pascal's law, which states that "Pressure applied to a liquid is transmitted equally in all directions," is very useful.

語　句

suddenly[sʌ́dnli] 副突然に　**pedal**[pédl] 名ペダル　**happen**[hǽpən] 動(事が)起こる→ happening 出来事　**system**[sístim] 名システム，組織　**receive**[risíːv] 動受け取る　**pressure**[préʃə] 名圧力　**transmit**[trænsmít] 動送る＝send　**movement**[múːvmənt] 名移動← move 動く　**immense**[iméns] 形広大な，ばくだいな　**liquid**[líkwid] 名形液体(の)　**exert**[igzéːrt] 動働かせる　**flat tire** パンクしたタイヤ　**inject**[indʒékt] 動注入する　**raise**[réiz] 動持ち上げる　**lift**[lift] 名リフト，エレベータ(英国で)　**device**[diváis] 名考案物，装置＝apparatus　**remote**[rimóut] 形遠い　**law**[lɔː] 名法律，規則，法則

●イントロ●　現在，パスカルといえばコンピュータ用の高水準言語で有名だが，液体の圧力に関するパスカルの原理もまた極めて有用である．このパスカルは同一人物で17世紀のフランスの科学者，哲学者でもあった．

▶**1. You all have seen**…➡現在完了の経験の用法．allは「あなた方は誰でも…を見たことがある」と強調している．

▶**2. must have stepped** ➡「きっと踏んだに違いない」．

▶**3. what happens** ➡ whatは先行詞と関係代名詞を兼ねていて，the accident which happensとしても同意．「起こった事件」．

▶**4. the liquid stored in the brake system, the brake oil**…➡ storedからsystemまではliquidを形容している．the brake oilとthe liquidとは同格．「ブレーキシステム中に蓄えられた液体，すなわちブレーキオイルは…」．

▶**4. great pressure, which is**…➡ whichの前に（,）があるので関係代名詞の継続用法．「そしてその大きな圧力は車輪に伝えられる」．

▶**8. the oil or air under pressure is conveniently used to**… ➡ 副詞（conveniently）はこのように助動詞（is）と本動詞（used）との間に入れて用いられる．「圧力のかかった油と空気は有効に…に用いられる」．

▶**9. "Pressure applied to a fluid is transmitted equally in all directions"** ➡（" "）は他人のことばなどを文中に引用するときに用いる符号で引用符（quotation mark）という．「液体に加えられた圧力はその強さを変えずにすべての方向へ伝達される」．

解 説　パスカルの原理の最も重要な応用は水圧（油圧）機械であろう．一般的に見られる油圧ジャッキ（hydraulic jack）の断面をFig. 3に示してある．ピストンAは小さい断面，Bは大きい断面をもっている．その両者は圧力伝達用の油で連絡されている．

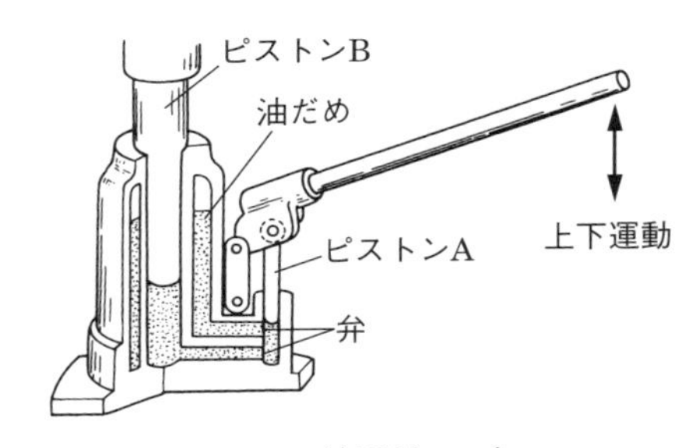

Fig. 3　油圧ジャッキ

ピストンAに加えられた圧力はそのままピストンBに伝送されるので，Bの断面積がAに比べて大きければBの面に発生する力は（Aの面積に反比例するから）極めて大きいものになる．したがって，ピストンAに加えた小さい力で大きな自動車を持ちあげるほどの力を発生することができる．

問題

1. 次の（　）の中の語を並べ変えて正しい英文とし，それを日本語に訳せ．
   - (1) You (into, inject, a flat tire, high pressure air).
   - (2) The driver (stepped, must, on his brake pedal, have).
   - (3) The movement (an immense pressure, is, of the car body, stopped, by).
2. 本文の内容と合っているものには○，合わないものには×をつけよ．
   - (1) The brake oil receives great pressure, which is transmitted to the wheel.
   - (2) Cars are easily raised by means of a lift operated by air under pressure.
   - (3) The oil under pressure conveys pressure different places.
   - (4) Pressure applied to a liquid is transmitted in one direction.

6 電気とは何か？

What Is Electricity?

1. Electricity exists in everything; the table, your body, clothing, and so on.
2. The reason we don't notice its existence is that it doesn't move.
3. Electricity at rest is called static electricity, and electricity in motion is called current electricity.
4. Electricity can do useful work only when it moves from one place to another.
5. Electricity is generated from batteries or generators.
6. In some ways, electricity is like water.
7. In the same way that water flows through a pipe, electricity flows along a solid wire.
8. Electricity gives us not only power, heat, and light, but also messages and entertainments on TV or radio.
9. Such materials as silver, copper, or aluminum are called conductors, because they readily conduct electricity from one place to another.
10. Such materials as plastic, rubber, air, or paper are insulators, because they have very poor conductivity.

語　句

exist[igzíst] 動存在する　**reason**[ríːzn] 名理由，根拠　**notice**[nóutis] 名動注意（する）　**existence**[igzístəns] 名存在←exist 存在する　**rest**[rest] 名休息，停止　**static electricity** 静電気　**battery**[bǽtəri] 名電池　**generator**[dʒénəreitə] 名発電機　**solid**[sɔ́lid] 名固体　**TV**＝television テレビ　**material**[mətíəriəl] 名物質，原料　**plastic**[plǽstik] 名プラスチック　**conductor**[kəndʌ́ktə] 名（熱・電気）の導体 → conductivity 伝導率　**insulator**[ínsjuleitə] 名絶縁体

●イントロ●　1でいろいろな形のエネルギーについて述べたが，なかでも電気は最も私達の生活に身近なものの一つである．

▶**1. everything; the table, your body…**➡語，句，節，文を区切るには主に（.）（:）（;）（,）が使われる．この順序で区切る力が弱くなり，それぞれ period（full stop），colon，semicolon，comma と呼ばれる．ここでは（;）は everything をさらに詳しく説明するための区切りに用いられている．

▶**4. only when it moves from one place to another**➡only は「唯一の」の意の副詞であるが，ここでは when…another の節を修飾するために文頭に置かれている．「それ（電気）がある点からほかの点へ移動するときにのみ」．

▶**7. In the same way that…**➡the same と that とは関連しており，「水がパイプの中を移動するのと同じように」．

▶**8. Electricity gives us not only…but also～**➡give は授与動詞なので間接目的語 us と，長い直接目的語 not only…but also～を必要とする．また not only と but also とは関連している．「…だけでなく～もまた」の意．

▶**9. Such materials as silver, copper…**➡この such と as とは関連しており例示を表す．「銀，銅，アルミニウムのような物質は導体と呼ばれる」．

▶**10. …, because～**➡このカンマ（,）は because を継続的につないでいる．「…は絶縁体である．その理由は～である」と訳す．

解説 **Conductor（導体）と Insulator（絶縁体）**➡両者の間に **semicondoctor（半導体）**といわれる一群の物質がある．シリコン（Si），ゲルマニウム（Ge），ガリウムひ素（GaAs）などで，それらは温度が上昇するにつれて，電子とホールとを発生し伝導性を増すようになる．この電子は金属導体が有する自由電子とは性質を異にするものである．それら半導体はまた微量の不純物が混入すると，電子あるいはホールを生じて著しく伝導性を増す．不純物を混入する操作を**ドーピング（doping）**という．またある半導体は光に当たると伝導性を増す．半導体はこのように普通の導体とは違った性質があるので，パソコンを動かす CPU，太陽電池，発光ダイオード，温度センサなどとして携帯電話，テレビ，洗濯機，エアコンなどさまざまなデジタル家電製品に使用され，我々の暮らしを支えている．

問 題

1. 次の（　）の中の語を並べ変えて正しい英文とし，それを日本語に訳せ．
 (1) Electricity (along, a copper wire, flows, water, like).
 (2) The conductor (electricity, conduct, to another, from one place, can).
 (3) The insulator (poor, very, has, conductivity).
2. 本文の内容と合っているものには○，合わないものには×をつけよ．
 (1) Electricity exists in your body.
 (2) Only generators can produce electricity.
 (3) Silver and copper are good insulators.
 (4) Rubber can be used for an electric wire.

コラム　コロンとセミコロンの使いどころ

コロン（colon, :）とセミコロン（semicolon, ;）は句読点の一種で，それぞれ異なる用法がある．文を区切る強さとしては，ピリオドとカンマの中間ぐらいである．コロンとセミコロンの右側にのみスペースを入れる．

コロン：文と文を繋げて，後者の文として，前文中の一部言換えや説明，引用文，選択肢などの例示の羅列などが導かれる．

セミコロン：複数の文を並列に繋げる際に用いられる．

以下は，コロンやセミコロンを用いた文章として，よく見かける文の構造である．

文 1 : 単語 a，単語 b，…，and 単語 z.

文 1 : 文 2a; 文 2b; …; and 文 2z.

7 電気回路

Electric Circuit

1. Electricity generated in one place will flow to another place through a path or circuit.
2. As shown in Fig. 4, electricity can do work, in this case lighting a lamp bulb, while it passes along the path.
3. Such a path usually makes a loop, and this loop is called an electric circuit.
4. If the switch is "off", the electric light fails. In this case, this circuit is called a broken circuit or open circuit.
5. When the light fails, electricity does not flow through the circuit.
6. A complete circuit or path is necessary to obtain useful work, and such a complete circuit is called a closed circuit.
7. When the electricity flows along an improperly closed path, it is called a "short circuit".
8. The short circuit sometimes causes unexpected dangers such as fire or electric shock.

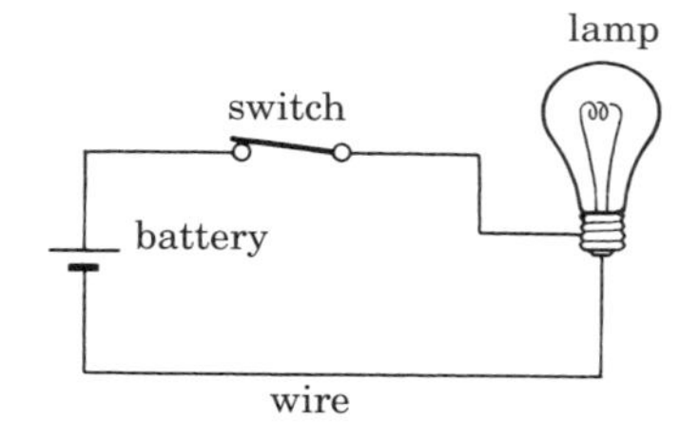

Fig. 4　Electric circuit

語　句

path[pæθ] 名経路，道　**circuit**[sə́:rkət] 名回路　**lamp bulb** 電球　**complete**[kəmplí:t] 形完全な，全部の　**improperly**[imprápərli] 副不適切に　**cause**[kɔ́:z] 動生じる，～の原因となる　**unexpected**[ʌ́nikspéktid] 形予期しない

●イントロ●　circuit（回路）とは電気の通る道筋で閉じているものをいう．もしも回路が不良であればさまざまな事故の原因となる．

構文

▶**1.　…will flow to another place through～**➡ through は「～を通って」の意なので，「…は道を通って別の場所へと流れていく」．

▶**2. …do work, in this case lighting a lamp bulb, while it passes alongthe path.** ➡ work「仕事」と lighting a lamp bulb「電球を点灯すること」とは同格.「それ（電気）が電線を流れている間に電球を点灯するという仕事をすることができる」の意.

▶**4. If the switch is "off", the electric light fails.** ➡ If…は仮定法といわれる用法.「もしもスイッチが切られているならば電灯はつかない」.

▶**7. along an improperly closed path** ➡副詞 improperly は closed を修飾し，closed は path を修飾している.「不適切な閉回路に沿って」.

▶**8. …causes unexpected dangers such as fire or electric shock.** ➡ dangers と such as 以下は同格.「…は火災とか感電とかの予期しない危険の原因となる」.

問題

1. 次の（ ）の中の語を並べ変えて正しい英文とし，それを日本語に訳せ.
   (1) Electricity (from one place, flows, to another).
   (2) An electric circuit (a closed circuit, be, should).
   (3) A closed circuit (to obtain, is, work , necessary).
2. 本文の内容と合っているものには○，合わないものには×をつけよ.
   (1) Electricity is able to light a lamp bulb.
   (2) A switch is used for turning electricity "on" or "off".
   (3) A short circuit is called an open circuit.
   (4) A short circuit is dangerous because it may cause a fire.

8 内燃機関

Internal Combustion Engine

1. An internal combustion engine, most commonly used in automobiles, consists of cylinders and pistons, and burns gasoline as fuel.
2. Internal combustion engines are roughly grouped into (1) reciprocating piston-type 4 cycle engines, (2) 2 cycle engines, and (3) rotary engines.
3. The motion of the 4 cycle engine is completed in four stages as follows : intake stroke, compression stroke, power stroke, and exhaust stroke, out of which only the power stroke produces energy to drive a car.
4. Thus energy is temporarily stored in a flywheel attached to the engine and then used to forcibly rotate the other three strokes.
5. The rotations per minute (rpm) of an ordinary automobile engine can vary from 1000 rpm to 5000 rpm.
6. This wide range of rpm in an engine is of great benefit for the maneuverability of automobiles, because cars are constantly requested to change speed.
7. The exhaust gas from the gasoline engines contains contaminating substances such as carbon monoxide (CO), nitrogen oxide (NO, NO_2, or the like), and not-yet-burned hydrocarbons (HC).
8. Before being exhaused, they are purified, while going through a muffler, by a chemical reaction on a three-way catalyst.

語 句

internal[intá:nal] 形内部の ⟷ external 外部の　**combustion**[kəmbʌ́stʃən] 名燃焼　**engine**[éndʒin] 名エンジン　**piston**[pístən] 名ピストン　**fuel**[fjú:əl] 名燃料　**reciprocate**[risíprəkeit] 動往復運動をする　**cycle**[saikl] 名循環，サイクル　**rotary**[róutəri] 動回転する　**stage**[steidʒ] 名段階，時期　**intake**[ínteik] 名取入れ（口）　**stroke**[strouk] 名一行程，ストローク　**compression**[kəmpréʃən] 名圧縮　**power stroke** 出力行程　**exhaust**[igzɔ́:st] 動排気する　**temporarily**[témpərərili]

副一時的に　**fly wheel** はずみ車　**rotations per minute** 1分間の回転数　**benefit**[bénifit] 名恩恵，利益　**maneuverability**[mənúːvəbíliti] 名操縦性　**exhaust gas** 排気ガス　**contaminate**[kəntǽmineit] 動汚染する　**carbon monoxide** 一酸化炭素　**nitrogen oxide** 酸化窒素　**not-yet-burned** 未燃焼の　**hydrocarbon** 炭化水素　**chemical**[kémikəl] 形化学の　**reaction**[riǽkʃən] 名反応 ← react 反応する　**catalyst**[kǽtəlist] 名触媒　**three-way catalyst** 三元触媒

●イントロ●　私達の現在の生活は自動車を抜きにしては考えられない．その心臓部であるエンジンについて述べている．

構文

▶**1. engine, most commonly used in automobiles,** ➡ most commonly used in automobiles は過去分詞による形容詞句で engine を修飾．「自動車で最も一般に使われているエンジン」．

▶**3. …is completed in four stages as follows :** ➡「次の四つの段階で完成する」．

▶**3. …, out of which only the power stroke produces energy** ➡わかりにくい構文だが，which の先行詞は四つの strokes.「四つの行程のうちで出力行程だけが自動車を駆動するエネルギーを供給する」．

▶**4. to forcibly rotate the other three strokes.** ➡「ほかの三つの行程を強制的に回転させるために」．このように副詞，forcibly を不完詞の to と動詞の間に割り込ませるのは文法的に正確ではないが，便利でわかりやすい表現なのでしばしば見られる．このような形を分割不定詞（split infinitive）という．

▶**7. contaminating substances such as…** ➡「…のような大気汚染物質」．contaminating は現在分詞による形容詞．

▶**8. by a chemical reaction on a three-way catalyst.** ➡「三元触媒上における化学反応によって」．三元触媒は排気ガス中の HC，CO，NO_x を無毒化するのに用いられる，白金を主体にした触媒．この触媒のおかげで大気汚染は大幅に減少した．

解説

発生した熱を暖房や調理に利用する以外に，熱エネルギーを動力に変換できるようになり，人間の労働が大幅に軽減されることになった．1700 年代後半に石炭火力による蒸気機関が完成し，産業革命がイギリスから始

まったが，人間に課せられていた苦役の解放革命ともいえる．しかし，この蒸気機関は熱を一度蒸気の形に変え，それから動力を得るしくみなのでその間の熱損失が大きく能率が悪い．ここで述べた内熱機関はシリンダ内で直接に熱料を燃やすので効率はずっとよくなる．

小型の内燃機関でどの位の発熱が有効に利用されるかというと，Fig. 5 に示されているように 25～28%に過ぎない．ずいぶん小さい値と思われるかもしれないが，これは熱力学の第二法則により一定の値以上にはできないことがわかっている．

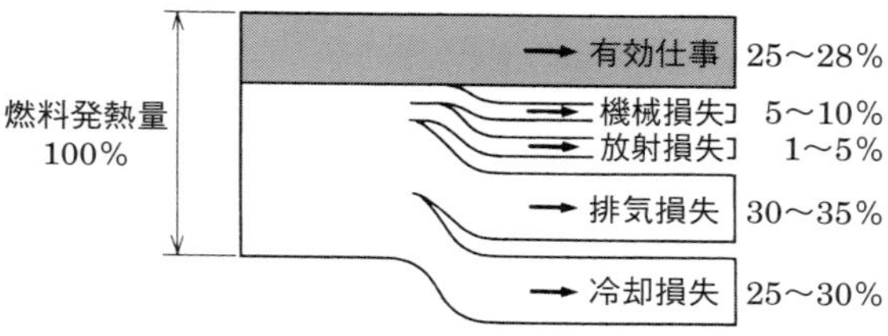

Fig. 5　内燃機関の熱の利用と損失

問 題

1. 次の（　）の中の語を並べ変えて正しい英文とし，それを日本語に訳せ．
   (1) An internal combustion engine (most, commonly, used, is, in automobiles).
   (2) An internal combustion engine (cylinders, and, consists, of, pistons).
   (3) The exhaust gas (contaminating, substances, contains).
2. 本文の内容と合っているものには○，合わないものには×をつけよ．
   (1) An internal combustion engine burns gasoline as fuel.
   (2) An internal combustion engine can be used for the electric cars.
   (3) A compression stroke is a stroke to produce energy in a 4 cycle engine.
   (4) A three-way catalyst is placed in the muffler to clean the exhaust gas.

9 メートル単位系

The Metric System

1. The metric system is a system used for measurement in scientific and engineering work in nearly all developed countries in the world.
2. This system is based on the decimal system in which each unit is a multiple or subdivision of 10.
3. The base units of the metric system are meter for length, kilogram for mass, and second for time.
4. If multiples or subdivisions of this unit are needed, we can obtain them merely by attaching a prefix to the base unit.
5. The common prefixes used for calculations are listed as follows.

Prefix	Value of prefix
milli-	1/1000
centi-	1/100
deci-	1/10
kilo-	1000

6. The chief advantage of the metric system is based on standards that have been accepted by international agreements.

語　句

metric[métrik] 形メートル(法)の← meter メートル　**measurement**[méʒəmənt] 名測定← measure 測る　**scientific**[sàiəntífik] 形科学の　**engineering**[endʒiníəring] 名工学← engine エンジン　**decimal**[désiməl] 形十進法の　**multiple**[mʌ́ltipl] 名倍数　**subdivision**[sʌ́bdiviʒèn] 名約数，細分　**mass**[mæs] 名質量　**prefix**[prí:fiks] 名接頭辞　**as follows** 次のとおり　**standard**[stǽndəd] 名標準　**international** [intənǽʃənəl] 形国際的な　**agreement**[əgrí:mənt] 名合意，一致

●イントロ●　計量の単位は私達の日常の生活はもちろん，学問の世界でも産業界でも極めて重要である．現在ではメートル，キログラム，秒を基にした**国際単**

位系（International System of Units）（SI 単位）が世界の主要な国で用いられている.

▶1. in nearly all developed countries in the world. ➡ nearly は almost とほぼ同意で「ほとんど」の意. almost は nearly よりも近接の意がある.「世界中のほとんどすべての先進国において」.

▶2. based on the decimal system in which each unit… ➡ based on は「…に基礎を置いて」. in which の先行詞は system であるがこの which は in の目的格になっている.「このシステムは, 各単位が 10 の何倍かまたは何分の一かになっている（いわゆる）十進法に基づいている」.

▶3. meter for length, kilogram for mass ➡この for は「…の代わりに」の意をもっているので,「長さ（の代り）にはメートルを用い, 質量にはキログラムを用いる」.

▶6. standards that have been accepted by… ➡ that は関係代名詞で, standards を先行詞とする. have been accepted は完了を意味する現在完了時制.「国際的に合意を得ている基準に基づいている」.

解説

単位（Unit）➡私達の日常の生活で“もの”を測ることは多い. A 地から B 地まで 100 km で, 自動車で行けば 2 時間かかる, などである. 測定値を表すには, ある基準の量の何倍かを用いることが多い. その基準の量を単位という.

単位は昔, 必要に応じて各国または各地方で単独で用いられていた. ところが交易や人の往来が盛んになるにつれて不便になり, 世界各国で共通の単位を制定する動きが起こった. その結果 18 世紀末に, フランスが主唱してメートル法による単位を世界各国で採用するに至った. しかし単位の切替えは国によってはいまだに十分には行われていない.

現在の**国際単位系**は 1960 年に決められたもので, 基本単位として, 長さにメートル〔m〕, 重さにキログラム〔kg〕, 時間に秒〔s〕, 電流にアンペア〔A〕, 温度にケルビン〔K〕, 物質量にモル〔mol〕, 光度にカンデラ〔cd〕を採用している. 面積などの単位は含まれていないが, 平方メートル〔m^2〕は長さ〔m〕から導かれる. 導かれたものを**組立単位（derived unit）**という.

問 題

1. 次の（　）の中の語を並べ変えて正しい英文とし，それを日本語に訳せ.
   (1) The metric system (a system, is, in almost every country, used).
   (2) The common prefixes (milli and kilo, for calculations, are, used).
   (3) 2000 meters (2 kilometers, usually, called, is).
2. 本文の内容と合っているものには○，合わないものには×をつけよ.
   (1) The metric system is based on the decimal system.
   (2) The metric system is used only in developed countries.
   (3) Meters, kilo-grams, and seconds are the base units of the metric system.
   (4) The advantage is that it is based on an international standard.

コラム by と with の違い

用いる手段や道具を示すときの前置詞「～で」「～を使って」として by と with があり，これらの使い分けに注意すること．方法・手順・プロセスなどの抽象的なものを表す場合は by を，道具・器具などの具体的な物質を示す場合は with が用いられる．移動手段の「バスで」を表す場合，by bus または with a bus となる．

Turbulent fluctuations were measured by particle image velocimetry.
（粒子画像速度測定法により乱流変動を測った.）
The wind speed was measured with a laser Doppler velocimeter.
（レーザードップラー速度計で風速を測った.）

10 測定と誤差

Measurements and Errors

1. The measurement of a physical quantity is always accompanied with a degree of uncertainty.
2. There are several reasons for the occurrence of measurement error.
3. Measuring instruments or devices have structural limitations restricting precise measurements.
4. The ambient conditions of measurement, for instance, temperature, humidity, and so on differ for each case.
5. The skill of the operator to handle instruments may differ from person to person.
6. Consequently, when measurements are reported, it is necessary to indicate the degree of uncertainty.
7. One way to express the uncertainty in a measurement is to use the "±" sign, as in 15.5±0.2 g.
8. If the weight of something is expressed as 15.5±0.2 g, this means its true weight definitely falls in the range from 15.3 to 15.7 g.

語　句

accompany[əkʌ́mpəni] 動伴う，同行する　**uncertainty**[ʌnsə́:tnti] 名不確実　**occurrence**[əkʌ́rəns] 名発生　**precise**[prisáis] 形正確な，精密な　**ambient**[ǽmbiənt] 形周りの，周囲の　**for instance** たとえば　**differ**[dífə] 形違う，異なる　**operator**[ɔ́pəreitə] 名運転者　**consequently**[kánsikwèntli] 副その結果　**indicate**[índikeit] 動指し示す，示す　**definitely**[défənətli] 副絶対に，確実に　**range**[reindʒ] 名一連，列，範囲

●イントロ●　測定の際，どんな精密な測定機械を用いても，また熟練した人が行っても必ず誤差が生ずる．誤差の原因および表示について述べている．

構文 ▶**8. its true weight definitely falls in the range from 15. 3 to 15. 7 g.** ➡「それの真の重さはまちがいなく 15.3 g から 15.7 g までの範囲に存在する」. range は「範囲」であるが，端から端までではなく「途中から途中まで」を意味する．

解説 数学で 165 と表せば，それは 165±0，つまり 165 ちょうどのことである．ところが測定値はそうではない．A 君の背の高さは 165 cm といった場合は 165±0 cm のことではなく，一般にはそれは 165±0.5 cm のことを意味している．測定値の場合には常に**誤差（error）**がつきまとうことを考えなくてはならない．

165±0.5 cm を数値線上に表すと Fig. 6 のようになる．つまり 164.5 ≦（背の高さ）＜ 165.5 cm のことである．

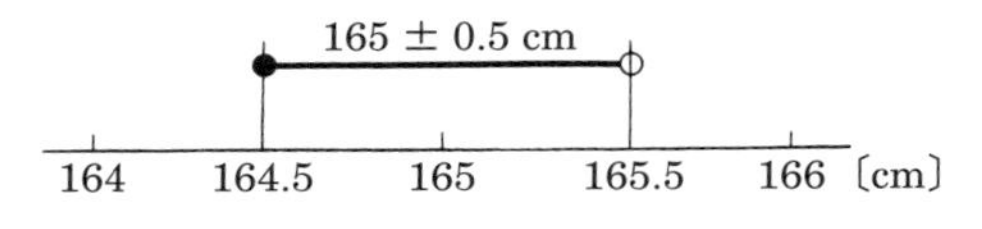

Fig. 6　165 ± 0.5 cm の数直線上の表示

問　題

1. 次の（　）の中の語を並べ変えて正しい英文とし，それを日本語に訳せ．
   （1）There (the measurement error, are, several reasons, for).
   （2）It (to indicate, necessary, the degree of uncertainty, is).
   （3）The "±" sign (used, in a measurement, to express, is, the uncertainty).
2. 本文の内容と合っているものには○，合わないものには×をつけよ．
   （1）The reports of measurements must indicate a degree of uncertainty.
   （2）The instruments have structural limitations restricting precise measurements.
   （3）Ambient conditions of measurements are always the same.
   （4）The skill of the operator differs from person to person.

11 有効数字

Significant Figures

1. In making measurement, record all integers that are certain, and one more in which there is some uncertainty.
2. Thus, all these integers are significant figures.
3. There is no relation, however, between the number of significant figures and the decimal point.
4. Each of 5.02, 50.2, 502, or 0.0502 clearly has three significant figures, but what about the number 5020?
5. In this case, we can't judge that the last zero may be significant or it may merely mark the decimal place.
6. To make it clear, the number should preferably be written as 5.02×10^3 or 5.020×10^3 using the power system（exponential notation）.
7. In this case, 5.02×10^3 means that three figures, 5, 0, and 2, are significant figures, and 5.020×10^3 has four significant figures 5, 0, 2, and 0.
8. Obviously, 5.020×10^3 is a more precise expression than 5.02×10^3, because the former has four significant figures and the latter only three.

語　句

significant[signífikənt] 形意味のある，有意の← sign 記号，サイン　**figure**[fígə] 名数字，図形　**integer**[íntedʒə] 名整数　**number**[nʌ́mbə] 名数(字)，番号　**whole number** 整数　**decimal**[désiməl] 形十進法の　**decimal point** 小数点　**judge**[dʒʌdʒ] 動判断する　**mark**[mɑ:k] 名記号，印 動示す　**should**[ʃud] 助…すべきである　**power system** 乗べき記法　**exponential notation** 指数記法(power system と同じ方法)　**obviously**[ɔ́bviəsli] 副はっきりと　**precise**[prisáis] 形正確な　**former**[fɔ́:mə] 形前の 代前者　**latter**[lǽtə] 形後の 代後者

●イントロ●　技術者や科学者は測定値を扱うことが多いので，その際，常に有効数字の概念をもつことが大切である．

▶**1. In making measurement,** ➡分詞構文をわかりやすくするために前置詞 in をつけたもの.「測定をするときには」の意.

▶**1. record all integers that are certain, and one more…** ➡命令文.「確実な数値全部と，ある程度不確実さのある数字を一つだけ記録せよ」の意. わかりにくい文だが，具体的に示すと，Fig. 7 のように mm 目盛りの物差しを用いて測定する場合に，33 mm までは確実に測れるが 33 mm と 34 mm の間は目測で測らなければならない. その目測による桁が不確実さをもっている桁ということになる. 今このブロックの横を 33.7 mm と読めば，33 が certain で 7 が uncertain な部分ということになる.

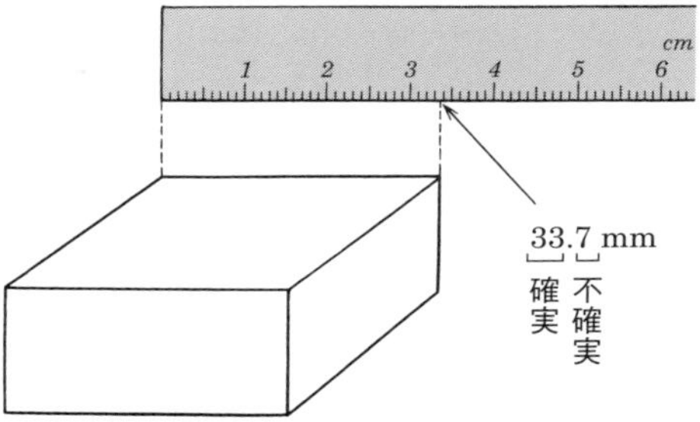

Fig. 7　ブロックの測定

▶**1. integers that are certain, 5. we can't judge that…** ➡二つの that は働きが違う点に注意. 前者の that は関係代名詞で「確実である全部の数字」の意. 後者の that は接続詞で that の前後の節を結びつける働きをしている.

▶**8. a more precise expression than 5.02×10^3** ➡形容詞 precise に more がついて比較級になったもの.「5.02×10^3 よりも正確な表現」.

▶**8. the former…the latter～** ➡対にして用いられることの多い語.「前者…，後者～」の意.

解説

「A 市の人口は 156000 人です」といった場合，常識的に約 156000 人だと考える. そのとき 1，5，6 の三つが有効数字であとの 0，0，0 は単に位取りを示すための零だと判断する. しかし自然科学ではこのようなあいまいさは許されない. どこまでが有効数字かを厳密に示す必要がある. そのために次の二つのどちらかが採用される.

(1) Exponential system（指数記法）　156000 人の代りに，1.56×10^5 とする. この場合 15.6×10^4 または 156×10^3 とはしない.

(2) Prefix system（接頭語法）　42200 m を有効数字 3 桁で 42.2 km とする.

問　題

1. 次の（　）の中の語を並べ変えて正しい英文とし，それを日本語に訳せ.
   （1） There is (no, the number of significant figures, relation, and, the decimal point, between).
   （2） The power system (used, of significant figures, is, to clarify, the number).
   （3） The number of significant figures of 5020 (judged, easily, can't, be).
2. 本文の内容と合っているものには○，合わないものには×をつけよ.
   （1） The number 5020 has four significant figures.
   （2） The number 5.020×10^3 has four significant figures.
   （3） 5.020×10^3 is a more precise expression than 5.02×10^3.
   （4） The number 5020 is not different from 5.02×10^3.

コラム and と or の使用

「A，B，C，とD」または「A，B，C，かD」の文はA，B，C，and DまたはA，B，C，or Dと書く．このときCのあとにカンマ（，）を置かない場合もあるが，置くことが望ましい．ただし項目が二つのときは（，）は不要である．このルールに従えば，一見読み難い「A，B and C，D，and E and F」と書かれた場合でも，「A」「BとC（のセット）」「D」「EとF（のセット）」の4組を並べたもの，と判断できる．

12 グラフ

Graphing

1. Various properties of substances are found to be interdependent; in other words, they have a functional relation.
2. When one value increases as the other decreases by the same factor, the relation is called an inverse proportion.
3. This can be expressed mathematically as $xy = a$ or $y = a/x$, where x and y: variables, and a: factor.
4. The relation between gas pressure and its volume is an example of an inverse proportion.
5. It is expressed as $PV = k$, where V stands for gas volume, P for gas pressure, and k for a factor of the inverse proportion.
6. The pressure and volume of 3.2g of oxygen gas are shown in the table below.
7. This relation can be plotted on a graph paper as shown in Fig. 8.

Table

Pressure〔kPa〕	Volume〔l〕
10.1	224
20.2	112
40.4	57.5
60.6	38.0
80.8	27.7
101.0	22.4

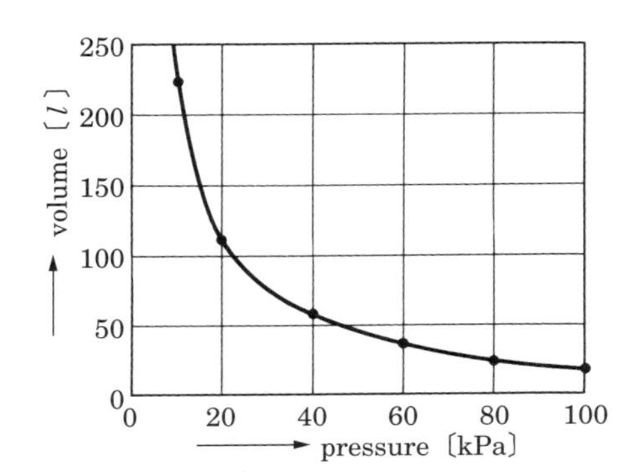

Fig. 8

語　句

plot[plɔt] 動図面を作る，グラフを描く　**property**[prɔ́pəti] 名特性，性質
substance[sʌ́bstəns] 名物質　**interdependent**[intədipéndənt] 形相互に依存する
functional[fʌ́ŋkʃənəl] 形機能の← function 働き　**increase**[inkrí:s] 動増大する
decrease[di:krí:s] 動減少する　**mathematically**[mæθimǽtikəli] 副数学的に
relationship[riléiʃənʃip] 名関係(していること)　**inverse**[invə́:s] 形逆の，反対の
stand for… を表す　**graph**[græf] 名動画グラフ，図表(を描く)

●イントロ●　数字や文書による記述は一次元的であって，ときに理解が困難である．それを二次元でみやすくするのがグラフである．グラフを簡潔にわかりやすく描くという習練は技術者にとって極めて重要である．

構文

▶**1.　…; in other words, they have a functional relation.** ➡ in other words「換言すれば」は i. e. とも書ける．「別のいい方をすればそれらは関数関係にある」の意.

▶**2.　inverse proportion** ➡「反比例または逆比例」．反対は正比例(directproportion)．

▶**5.　It is expressed as *PV*=*k*, where *V* stands for gas volume, *P* for gas pressure,…** ➡ where 以下は $PV=k$ の式中の文字（P や V）が何を表しているかを説明する部分で「ここで，V は…，P は…」と訳す．「ここで，V は気体の体積，P は気体の圧力，…を表す」．

▶**6.　3.2 g of oxygen gas** ➡「酸素ガスの 3.2 g」．of は省略可．単位のついた数値はそのまま形容詞として用いることができる．The rod of 10 mm diameter is used here.「10 mm 直径の棒がここで使われている」．

▶**Fig. 8** ➡ graph では横軸（abscissa）と縦軸（ordinate）の選択を誤らないように注意する．横軸には原則として温度，圧力，時間などを取る．

解説　グラフには本文に示した純粋に数学的なものから，わかりやすいように形を変えたものまでいろいろある．

問　題

1. 次の（　）の中の語を並べ変えて正しい英文とし，それを日本語に訳せ.
   (1) A functional relation (found, of substances, in various properties, is).
   (2) The relation (is, between, its volume, an inverse proportion, and, gas pressure).
   (3) An inverse proportion ($y=a/x$, expressed, mathematically, is, as).
2. 本文の内容と合っているものには○，合わないものには×をつけよ.
   (1) There is no realtion between gas pressure and its volume.
   (2) The relation between gas pressure and its volume is a direct proportion.
   (3) "$y=a/x$" is an example of an inverse proportion.
   (4) An inverse proportion is always plotted as a straight line on a graph paper.

コラム　ラテン語の略語

科学論文などの技術英文では，ラテン語の略語が使われることがある.

e.g. (exempli gratia) = for example
etc (et cetera) = and other things
et al. (et alii) = and others
i.e. (id est) = that is to say
cf. (confer) = confer, refer

○● 章末問題 ●○

1．空欄を埋めて，日本語に訳せ．

(1)　Motors change electrical energy (　　　) mechanical energy.

(2)　The coin may fall faster (　　　) the feather.

(3)　These phenomena are all caused (　　　) friction.

(4)　It is thought to work (　　　) a bent lever.

(5)　The driver (　　　) have stepped on his brake pedal.

(6)　Electricity is generated (　　　) batteries or generators.

(7)　Electricity flow (　　　) the circuit.

(8)　The rpm of an ordinary automobile engine can vary (　　　) 1000 rpm (　　　) 5000 rpm.

(9)　This system is based (　　　) the decimal system.

(10)　The skill may differ (　　　) person (　　　) person.

(11)　Record all integers (　　　) are certain.

(12)　There is a functional relation (　　　) gas pressure (　　　) its volume.

2．次の文を英語に直せ．

(1)　エネルギーはいろいろな形で存在する．

(2)　羽とコインをつかみ，同じ高さから落とせ．

(3)　昔，人々は火をおこすため木の棒をこすりあわせた．

(4)　私たちは釘にかかる力 x を計算することができる．

(5)　あなた方みんな，車が突然止まったのを見たことがある．

(6)　私たちがその存在に気が付かない理由は，それが動かないからです．

(7)　電気が不適切な閉回路に沿って流れるとき，それをショートした回路と呼ぶ．

(8)　ガソリンエンジンからの排気ガスは大気汚染物質を含む．

(9)　メートル単位系は科学と工学の仕事で測定のため使われる単位系です．

(10)　物理量の測定はいつもある程度の不確かさを伴う．

(11)　有効数字の数と小数点の間には何の関係もない．

(12)　これは数学的に $xy=a$ と表すことができる．

II

機械工学の周辺

Periphery of Mechanical Engineering

機械工場で，図面とともに業務命令を受け取った場合，直ちに仕事を始めるというわけにはいかない．まず図面を詳しく読み，材料を選択し，加工の方法や順序を考え，さらに工員を配置するなど，数多くの段取りが必要とされる．

この章では，そのような段取り（仕事の手順）に相当する内容を中心に述べている．

職場ではよく，「段取り半分」ということがいわれる．段取りができあがれば仕事は半分終わったも同然だ，という意味である．何事によらず，計画を途中で変更することのないようにしなければいけない．

1 製作図

Working Drawing

To make an object according to required specifications, the worker must have information on its precise shape and dimensions, special processes needed to make, materials to be used, and finish. From a pictorial drawing, as shown in Fig. 9, it is usually difficult to obtain the detailed information needed. In such cases, we require two or three projections by which we can obtain the exact shape and dimensions. Such working drawings should satisfy all requirements and be completely clear and understandable.

For that purpose, at least two views are used, and if necessary, three or more views are used. If we try to draw the block shown in Fig. 9 with a three-view system, we can get Fig. 10. Three views show three different surfaces of the block ; that is a front view, top view, and side view.

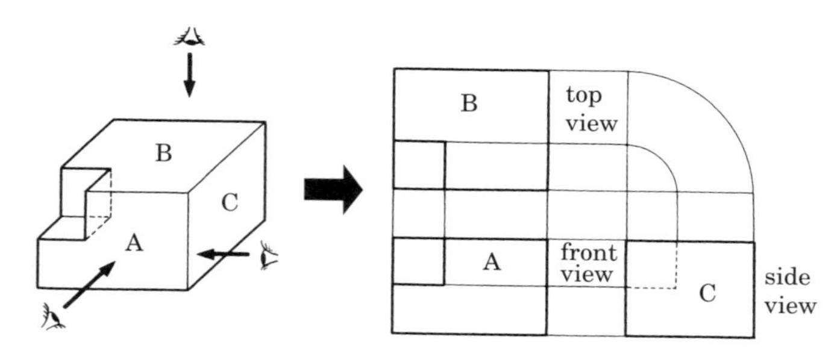

Fig. 9　Pictorial drawing

Fig. 10　Three-view drawing

語　句

drawing[dró:iŋ] 名図面，製図← draw 製図する，引く　**object**[ɔ́bdʒikt] 名物体，目的　**according to**…に従って　**specification**[spèsifikéiʃən] 名仕様書← specify 詳述する　**information**[ìnfəméiʃən] 名知識資料← inform 知らせる　**shape**[ʃeip] 名動形状(を作る)　**dimension**[diménʃən] 名寸法＝size　**pictorial drawing** 見取図　**detail**[dí:teil] 名動詳細(に述べる)　**projection**[prədʒékʃən] 名投影図(法)　**satisfy**[sǽtisfai] 動満足させる　**purpose**[pə́:pəs] 名目的＝**aim**　**view**[vju:] 名画面，光景　**three-view system** 三面図法　**surface**[sə́:fis] 名表面　**front view** 立面図　**top view** 平面図　**side view** 側面図

●イントロ●　機械工作の出発点である図面，三面図について述べている．例示した投影図は直角投影図法のうち第三角法によるものである．

▶**information on its precise shape…, special processes…, materials…**➡「その正確な形状と寸法，特殊な工作，材料，および仕上げについての知識をもたねばならない」．カンマ（,）で区切られた四つの部分は同格で on に関係している．最後は and でつなぐ．

▶**it is usually difficult to obtain…**➡ it は to 以下の節の代わりに用いられている形式主語．「…を得ることは通常は困難である」．

▶**projections by which we can obtain…**➡ which の先行詞は projections.「正確な形状や寸法を得ることができるような 2 枚または 3 枚の投影図を必要とする」．

▶**…should satisfy all requirements and be completely clear**➡「…は要望を満たし完全に明瞭でなければならない」．satisfy と be は同じ関係で should につながる．

▶**three different surfaces of the block; that is …**➡ three different surfaces と that is 以下は同格の用法．「立体の三つの面を示す．すなわち立面図，平面図，および側面図である」．

解 説 工作図は機械工作の基礎として重要なものなので，その図面は誰が見ても同じように正しく判断できるものでなくてはならない．そのため，図面の描き方，すなわち図法が日本では**JIS**（ジス）で厳密に決められている．図法が一見数学のようなかた苦しい印象を与えるのはそのためである．簡単な工作物であれば，Fig. 9 に示されるような pictorial drawing（見取図）でも理解されるが複雑なものになるとどうしても Fig. 10 に示されるような直角投影図によらなければならない．

なお，JIS は**日本工業規格（Japanese Industrial Standards）**の略で，日本の工業製品の品質，形状，検査法などについて規格を定めたものである．JIS マーク表示認定工場で作られた品物は一定以上の品質が保証されている．

問 題

1. 次の（　）の中の語を並べ変えて正しい英文とし，それを日本語に訳せ．
   （1） The worker (and, on, must, information, its precise shape, have, dimensions).
   （2） A pictorial drawing (the exact shape, suitable, to obtain, and, is, not, dimensions).
   （3） We (to be made, require, two projections, on the object, at least).
   （4） The working drawings (clear, must, and, be, understandable).
2. 本文の内容と合っているものには○，合わないものには×をつけよ．
   （1） We can get sufficient information from a pictorial drawing.
   （2） It is difficult to obtain the exact shape and dimensions from a pictorial drawing.
   （3） Two views are enough to obtain the exact shape and dimensions.
   （4） Three views are preferable to obtain detailed information.

2 工作図の線

Lines for Working Drawing

Several kinds of lines are used in a working drawing. Beginners must practice to make sure that correct lines are used in proper places.

A visible line is represented by a continuous thick line, usually having a thickness of 0.6 to 0.8 mm, used to show all edges that can be seen. A hidden line, sometimes called an invisible line, is composed of thick dashes about 3 mm long alternated with about 1 mm spaces. This dashed line is employed to show hidden edges which cannot be seen from the perspective of the drawing.

An extension line is a thin line drawn from the edge of the object from which measurements are to be made. A dimension line is also a thin line, which is drawn between two extension lines and has arrow marks at both ends. The figure of dimension is centrally written on the dimension line, as shown in Fig. 12.

A break line is the one by which the object is imaginably cut to show the inside condition of the object. This line is usually a thick wavy line.

Whenever we talk about the distance from one hole to another, we mean from the center of the hole to the center of another, in other words, center-to-center distance.

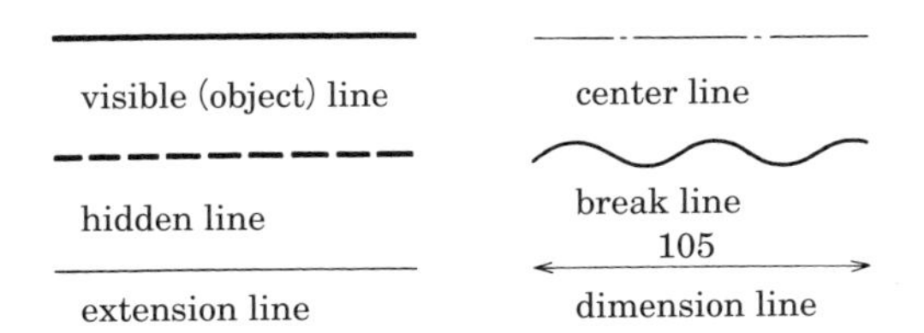

Fig. 11 Lines

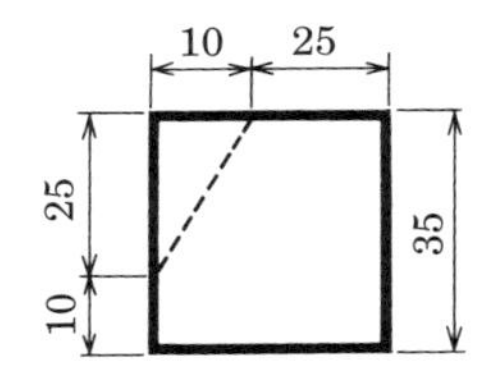

Fig. 12 Dimensioning

語　句

beginner [bigínə] 名初心者← begin 始める　**practice** [prǽktis] 名練習　**correct** [kərékt] 名正しい　**proper**[prɔ́pə] 名適当な　**visible**[vízəbl] 形目に見える→ vision 視覚，洞察力　**continuous**[kəntínjuəs] 形連続的な　**continuous thick line** 太い実線　**edge** [edʒ] 名境，へり　**hidden** [hídn] 形隠された← hide 隠す　**invisible** [invízəbl] 形目に見えない　**compose**[kəmpóuz] 動…から成る（of）　**alternate** [ɔ́:ltənit] 動交互にする　**dashed line** 破線　**extension**[iksténʃən] 名延長← extend 延長する　**thin line** 細線　**dimension line** 寸法線　**opening**[óupəniŋ] 名空所，オープニング　**break line** 破断線　**imaginably**[imǽdʒinəbli] 副想像できるように　**wavy**[wéivi] 形波のような　**whenever**[hwenévə] 副　いつでも　**distance** [dístəns] 名距離　**center-to-center** 中心から中心まで

●イントロ●　製図法で用いられる線について述べている．

▶**Several kinds of lines** ➡線の種類がいくつかある場合，このように kinds と lines と共に複類にする．lines of several kinds ということもある．「何種類かの線」の意．

▶**a continuous thick line, usually having…, used to show…,** ➡少し複雑な構文であるが a continuous thick line が二つの形容詞句 having…, と used to …とで形容されている形．「太さが 0.6 から 0.8 mm であり，そして外形を示すために用いられる，連続している太い線」．

▶**dashes about 3 mm long alternated with…** ➡ alternated 以下で dash を形容している．「約 1 mm の間隔を交互にもっている 3 mm の太いダッシュ線」とは破線（dashed line）の意．点線とは違うので注意．

▶**the object from which measurements…, と a thin line, which is…,** ➡この二つの which は，その前のカンマ（,）の有無によって訳し方が異なる．前者は「そこから測定が始まる物体の端」，後者は「…は細線であり，その細線は…の間に引かれる」の意．

▶**whenever we talk about…** ➡「私達が…について語るときはいつでも」．

問題

1. 次の（　）の中の語を並べ変えて正しい英文とし，それを日本語に訳せ．

（1）A visible line (line, a continuous, is, straight, thick).

（2）A dimension line (arrow marks, has, both, at, ends).

（3）A hidden line (a thick line, composed, is, dashes, of, and, spaces).

（4）An extension line (thin, a continuous, is, straight, line).

2. 本文の内容と合っているものには○，合わないものには×をつけよ．

（1）A continuous thick wavy line is a break line.

（2）A break line is the line where the object is cut.

（3）A thin line with a figure on it is a dimension line.

（4）The distance from one hole to another means distance between edges.

コラム　must と have to の違い

義務「～しなければならない」を表す際に，must または have to が用いられるが，これらもニュアンスはまったく異なる．will と be going to の場合と同様（→ p. 11），前者は助動詞であり，書き手・話し手の主観的な気持ちが内在する．つまり，must では，内から（本人から）課される義務を言い表し，積極性を感じさせる．一方，have to は客観的事実，または他者から課される義務を述べる際に用いられる．

We must stop smoking, because our health should be at risk.
（健康被害の恐れがあるので，禁煙しなければならない．）

We have to stop smoking, because this restaurant is smoke-free.
（このレストランは禁煙のため，タバコは止めなければならない．）

3 マイクロメータの取扱い

Care for a Micrometer

A micrometer is a precision instrument designed to measure small distances or shallow hollows. This instrument must be handled as carefully as possible. Dropping it on the floor or bench may damage its fine parts, rendering it useless. It should be kept free from dust, grit, and grease. The spindle should not be in contact with the anvil when not in use, because temperature change may put unnecessary strain on the spindle.

The spindle of a micrometer is an accurately machined screw, and it is rotated by a thimble or ratchet knob until the object to be measured is in contact with both the spindle and anvil. The ratchet ensures correct pressure between the anvil and spindle due to its ratchet mechanism. Each graduation on the sleeve represents 0.5 mm, which is subdivided into 0.01 mm on the thimble scale. A vernier scale allows accurate reading to 0.001 mm. Its exactness should be tested now and then using a gauge block.

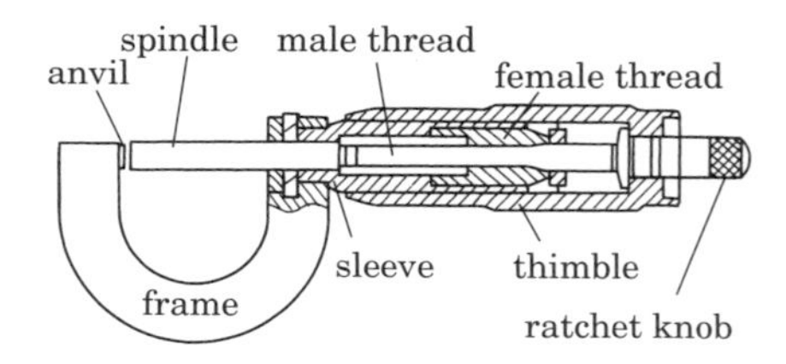

Fig. 13　Micrometer

語　句

micrometer[maikrɔ́mitə] 名マイクロメータ　**instrument**[ínstrumənt] 名器具，道具　**hollow**[hɔ́lou] 名くぼみ，穴　**damage**[dǽmidʒ] 名動損害(を与える)　**render**[réndə] 動(ある状態)にする，提出する　**free from**… の存在しない　**temperature**[témprətʃə] 名温度　**unnecessary**[unnésisəri] 形不必要な　**strain**[strein] 名ひずみ　**spindle**[spíndl] 名軸，心棒　**machine**[məʃí:n] 名動機械(加工する)　**rotate**[routéit] 動回転する→ rotation 回転　**ratchet**[rǽtʃit] 名つめ車，ラチェット　**knob**[nɔb] 名つまみ，ノブ　**thimble**[θímbl] 名シンブル　**anvil**[ǽnvil] 名アンビル　**graduation**[grædjuéiʃən] 名目盛り← graduate 目盛りをつける　**sleeve**

[sli:v] 名スリーブ，袖　**vernier**[vә́:rjә] 名バーニヤ（→解説）　**exactness**[igzǽktnis] 名正確，精密　**now and then** ときどき　**gauge block** ブロックゲージ

●イントロ●　どこの機械工場でも見られる，極く普通の精密測定工具であるマイクロメータの取扱いを述べている．

▶**as carefully as possible** ➡「できるだけ注意深く」．
▶**…should not be in contact with the anvil when～** ➡ この when は接続詞．「使わないときは…をアンビルと密着させてはいけない」．
▶**due to its ratchet mechanism.** ➡「ラチェット機構を利用して」．ラチェット機構とは，ねじで物を締めつけるときに，ある圧力以上になるとねじが空まわりしてそれ以上圧力がかからないようにする保護機構をいう．
▶**A vernier scale allows accurate reading to 0.001 mm.** ➡「バーニヤ尺を利用すれば 0.001 mm の精度まで読み取ることができる」の意．

解 説　バーニヤ尺，副尺（vernier scale）Fig. 14 において本尺の 9 目盛りを副尺では 10 等分してある．図のように丸棒をはさんで測定するときに，本尺と一致している副尺の目盛り（図では 7 の点）を読めば，本尺の 1 目盛りの 1/10 まで正確に読み取ることが可能になる．Fig. 14 の丸棒の径は 0.7 mm と読める．本尺の 11 目盛りを 10 等分した副尺ではどうなるだろうか．厚紙で本尺と副尺を作り，試してみよ．

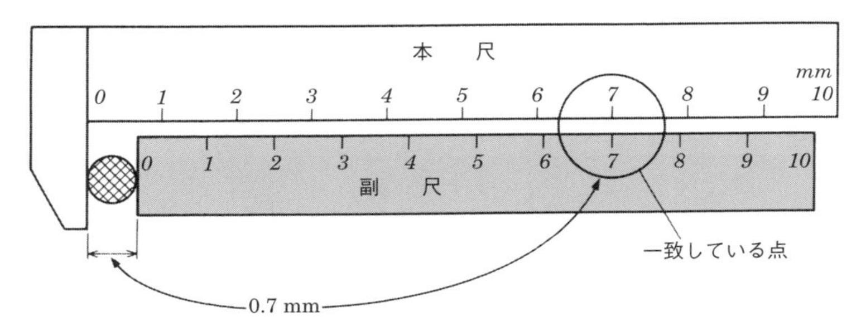

Fig. 14　バーニヤ尺の原理図

問　題

1. 次の（　）の中の語を並べ変えて正しい英文とし，それを日本語に訳せ．

1. (1)　A micrometer (is,　small distances,　a precision instrument,　to measure).
   (2)　A micrometer (be,　carefully,　handled,　as,　must,　as possible).
   (3)　A micrometer (dropping,　may,　damaged,　by,　be,　it,　on the floor).
   (4)　A micrometer (be,　contamination,　in the case,　stored,　must,　to prevent).

2. 本文の内容と合っているものには○，合わないものには×をつけよ．
   (1)　You have to know how to use a micrometer.
   (2)　The object is properly placed between the spindle and anvil of a micrometer.
   (3)　You have to rotate a thimble of a micrometer to measure an object.
   (4)　We don't need to check the exactness of a micrometer.

コラム　another と the other の違い

「その他の～」として another と（the）other があり，これらには明確な違いがある．another は「an ＋ other」から成るため，「その他のある一つの～」を意味し，another ＋ 単数名詞となる．another の前に the が付くことはない．「その他のあるいくつかの～」の場合は other（＋ 複数名詞）を，「その他のすべての～」の場合は the other（＋ 単数/複数名詞）となる．

4 ゲージ類

Gauges

A gauge is a device to determine the size of an object. There are many types of gauges, and they can be categorized roughly into two types: gauges such as a micrometer or vernier caliper, which are graduated, are included in one group; and the other group includes devices called end standards, which have no graduation but give very accurate dimension from one side to another. A typical end standard is a gauge block or a limit gauge.

When calibrating a graduated gauge or a limit gauge, we use a gauge block, which is a hardened steel block finished extremely precisely. Blocks are sold in a set so that any size desired for inspection can be produced by combining several blocks. The best known gauge block was invented by Carl E. Johnson in Sweden in 1855, and it is called Johnson gauge or Jo-gauge for short.

You may wonder how Jo-gauge was orginally calibrated. The final standard of length depends on the wavelength of red light at the end of the spectrum. The dimension of a gauge block is determined by an interferometer to the nearest 25 nm. Gauge blocks translate this basic linear standard into practical form for shop use.

語 句

categorize[kǽtigəraiz] 動類別する← category 種類　**roughly**[rʌ́fli] 副荒く，粗雑に ← rough 荒っぽい　**vernier caliper** ノギス　**end standard** 端度器　**graduation**[grædjuéiʃən] 名目盛り　**typical**[típikl] 形模範的な← type 典型，模範　**limit gauge** 限界ゲージ　**calibrate**[kǽlibreit] 動目盛り校正をする　**harden** [há:dn] 動硬くする← hard 硬い　**inspection**[inspékʃən] 名検査← inspect 検査する　**combine**[kembáin] 動結合させる　**originally**[ərídʒinəli] 副最初は　**wavelength** [weivléŋkθ] 名波長　**spectrum**[spéktrəm] 名スペクトル　**interferometer** [intəferɔ́mitə] 名（光の）干渉計　**translate**[trænsléit] 動翻訳する　**practical** [prǽktikl] 形実際的な

●イントロ●　ゲージブロック（日本ではブロックゲージ）は精密測定の基礎になるゲージで，ヨハンソンゲージが有名である．

構文　▶**vernier caliper, which are graduated, are included** ➡ which are graduated の前後にあるカンマは継続用法のためのものではなく（　　）の働きをもつ．「目盛りのついているノギス…などは一つのグループに含める」．

▶**end standard, which have no graduation** ➡この which は関係代名詞の継続用法で，「その端度器は目盛りをもっていないが，極めて精密な測定値を与える」の意．

▶**Blocks are sold in a set so that any size desired…** ➡理由を示す so that は，後の can と関連している．「検査に必要などんな寸法でも作れるようにセットで売られている」．

▶**The best known gauge blocks was…** ➡ best は副詞の最上級の用法．「最もよく知られているブロックゲージは…によって発明された」．

▶**The final standard…depends on the wavelength** ➡ depends は on または upon に関連して用いられており，「…に左右される」の意．「長さの最終の基準は…の波長に依っている」．

▶**to the nearest 25 nm** ➡「25 nm の精度で」．この用法は便利．例えば，「0.1 mm の精度まで測れ」は Measure it to the nearest 0.1 mm. と書ける．25 nm は 0.000000025 m の長さ．“n” は nano の略で，10^{-9} の意．

問題

1. 次の（　）の中の語を並べ変えて正しい英文とし，それを日本語に訳せ．
   (1)　A gauge (to determine,　the size,　is,　a device,　of an object).
   (2)　A gauge block (the typical end standard,　is,　of,　one).
   (3)　Jo-gauge (used,　calibrating,　is,　a graduated gauge,　for).
   (4)　The final standard of length (red light,　the wave length,　is,　of).
2. 本文の内容と合っているものには○，合わないものには×をつけよ．
   (1)　There are two types of gauges.

(2)　We need several gauge blocks to calibrate a micrometer.
(3)　Jo-gauge was orginally calibrated by the other gauge blocks.
(4)　A micrometer is usually more precice than a gauge block.

コラム　could は「～できた“かもしれない”」

「～できた」「～することができた」を表すとき，日本人の陥りやすい誤訳として can の過去形 could を使いがちである．しかし，ネイティブの受取り方は，低い可能性を意図したと判断して「～できるかもしれない」，もしくは仮定法の一部と判断して「～できたかもしれないけど，できていない」と勘違いされる．日本語での「過去に実現できた」ことを伝えるには，could は不要であり，過去の事実を伝える単純な過去形で十分である．例えば，「成功した」というポジティブな意図を含めたければ successfully を付ければよい．

We could demonstrate the feasibility.（実現可能性を実証できたかもしれない.）
We demonstrated the feasibility.（実現可能性を実証できた.）
We successfully demonstrated the feasibility.（実現可能性を上手く実証できた.）

5 限界ゲージ

Limit Gauge

A limit gauge is a double gauge of which one side is the "go" side and the other side is the "no go" side. It can indicate whether or not the dimension of a part falls between the upper limit and lower limit previously specified on the design drawing, hence the name limit gauge was given. If the workpiece can pass through the large dimension, the upper limit, and is stopped by the small dimension, the lower limit, then the dimension of the workpiece lies between these two limits, and the object is judged to be of proper dimension for the functional purpose. If not, the part should be discarded or rejected.

Each gauge is stamped with its size. One is usually stamped "Go," while the other is stamped "No go." Limit gauges are, therefore, also known as "go and no go gauge."

For line production and inspection work, this limit gauge system provides a rapid and accurate means of dimension measurement and control. This system has dramatically contributed to mass production by helping to realize the concept of interchangeability.

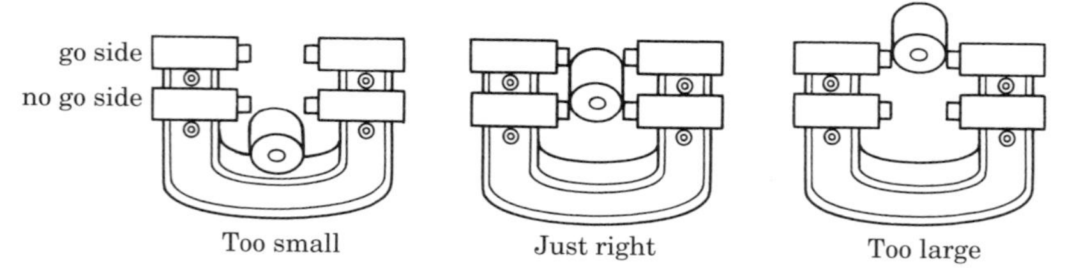

Fig. 15　limit gauge

語　句

fall[fɔ:l] 動落ちる，倒れる，ある　**previously**[prí:viəsli] 副以前に，前もって　**hence** [hens] 副それゆえに，したがって　**discard**[diská;d] 動捨てる　**stamp**[stæmp] 名動 型，印(を押す)　**dramatically**[dramǽtikəli] 副劇的に　**mass production** 大量生産

●イントロ● 限界ゲージによる検査とは，わずかに寸法差のある二つのゲージを作り，片方を通過して他方を通過しない部品を合格とする方法をいう．近代大量生産方式の基礎を作った検査方式である．

構文 ▶**It can indicate whether or not the dimension of a part falls between…and〜**➡ whether or not…「…であるかどうか」．「部品の寸法が…と〜との間にあるかどうかを示すことができる」．この or not は省略することも，もっと後に置くこともできる．

▶**large dimension, the upper limit,**➡同格の用法．「大きい寸法，すなわち上限…」．

▶**This system has dramatically contributed to…**「このシステムは…に対して劇的な大成功をおさめた」．現在完了の継続の用法．

問 題

1. 次の（　）の中の語を並べ変えて正しい英文とし，それを日本語に訳せ．
   (1) A limit gauge (of the workpiece, is, the size, a gauge, to test).
   (2) The size of the workpiece (the lower limit, be, should, the upper limit, and, between).
   (3) A limit gauge system (to, mass manufacturing, is, essential).
   (4) A limit gauge system (relation, a close, has, to interchangeability).
2. 本文の内容と合っているものには○，合わないものには×をつけよ．
   (1) A limit gauge has graduated scales.
   (2) A limit gauge is a sort of end standard.
   (3) A limit gauge usually has two sides, a “go” side and a “no go” side.
   (4) A limit gauge can be used for the calibration of a gauge block.

6 応力ひずみ線図

Stress-Strain Diagram

When we are planning to construct a building or machine with sufficient safety factor, it is very important to know the limit of endurance of materials to exterior forces or stress.

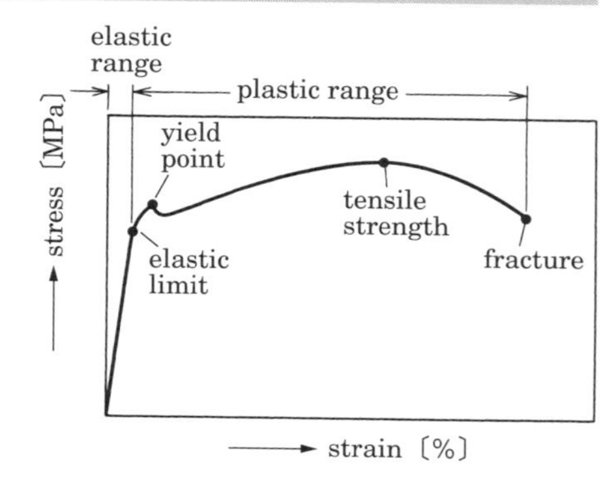

Fig. 16　Stress-strain diagram of low-carbon steel

The stress-strain diagram is one way to show the relation between stress (determined by dividing the load〔Pa〕by original area of cross section〔cm^2〕) and strain〔%〕(obtained by dividing increment of gauge distance〔cm〕by original length〔cm〕). Fig. 16 shows a typical stress-strain diagram of low carbon steel.

In this diagram, the straight line from 0 to elastic limit is called the elastic range where the strain is directly proportional to stress, and the factor of this relation is called the modulus of elasticity (Young's modulus).

When stress increases beyond the limit, in other words, when stress exceeds the elasticity limit, the object cannot recover its original shape, even if the stress is removed. The point where strain increases without an increment of stress is called a yield point and the stress given at the yield point is used as a base for planning construction.

The tensile strength is obtained by dividing the maximum stress by the original cross-sectional area.

語　句

sufficient[səfíʃənt] 形十分な　**safety**[séifti] 名安全← safe 安全な　**factor**[fǽktə] 名要素，係数　**endurance**[indjúərəns] 名忍耐(力)　**exterior**[ekstíəriə] 形外部の

⟷ interior 内部の　**stress**[stres] 名圧力，応力，ストレス　**load**[loud] 名負荷　**stress-strain diagram** 応力ひずみ線図　**cross section** 断面　**increment** [ínkrimənt] 名増大　**low-carbon steel** 低炭素鋼　**diagram**[dáiəgræm] 名図表，ダイヤ　**elastic**[ilǽstik] 形弾性の（ある）　**modulus of elasticity** 弾性係数　**yield point** 降伏点　**tensile strength** 引張強さ

●イントロ●　応力ひずみ線図は構造設計などの基礎となるもので，ここでは線図の構造，用途などについて学ぶ．

構文　▶**When we are planning…, it is very important to know～**➡「私達が…の設計をするときに，～を知ることは極めて重要である」．when は接続詞．it は to 以下を表す形式主語．

▶**determined by dividing the load〔Pa〕by original area**➡「荷重を元の面積で割ることで計算できる」．divide A by B「A を B で割る」．multiply A by B「A に B をかける」．

▶**the elastic range where the strain is directly proportional to stress**
➡ where は関係副詞の限定用法．「ひずみが応力に比例するような弾性の範囲」．

▶**the object cannot recover its original shape, even if the stress is removed.**➡「たとえ応力を取り除いても，その物体は元の形状を回復することはできない」．

解説　素材を使ってものを作るときに素材の性質を知る必要がある．そのための試験を材料試験といい，いろいろな方法がある．硬さ試験とここで述べた引張試験が主なものである．材料を強く引張れば，どんな材料でも必ず切れるが，その限度には大きな違いがある．引張強さの大きい材料を使えば，建造物を軽量にすることができ，経費を安くすることができる．外力が小さいうちはそのひずみは外力に比例し，外力を取り去れば，試験片はもとの形に戻る．この範囲は図の直線部分であって，**elastic range（弾性範囲）**といわれる．それ以上に大きい力をかけると，外力を取り去っても元の形に戻らず，やがては切断する．この範囲を **plastic range（塑性範囲）**という．

問　題

1. 次の（　）の中の語を並べ変えて正しい英文とし，それを日本語に訳せ．

(1) It is important (of endurance, to exterior forces, the limit, to know, of materials).

(2) A stress is obtained (of the cross section, by, the load, by, the original area, dividing).

(3) An elastic range is the range (is, the strain, proportional, where, to the stress).

(4) A yield point is the point (of stress, increases, without, strain, an increment, where).

2. 本文の内容と合っているものには○，合わないものには×をつけよ．

(1) The stress-strain curve represents the relation between stress and strain.

(2) The elastic limit is the same as the tensile strength.

(3) The object can't recover its original shape if the stress is over the elastic limit.

(4) The tensile strength is used as a base for planning construction.

コラム　括弧の使いどころ

文中の括弧は追加情報を挿入する際に用いられる．特に，生誕年や発表年を記すときによく使用される．「Albert Einstein (1879–1955)」「a pioneering study of Reynolds (1883)」．このとき括弧の前後にスペースを入れ，括弧の内側にはスペースは不要である．カンマやエムダッシュの挿入表現とは異なり，必ずしも不可欠ではない情報や本文とは関連性の弱い情報を挿入する際に用いられる．つまり，和文ではほとんど意識されていない括弧の用法であるが，英文では読者に読み飛ばされても構わない場合にのみ用いるよう，注意すること．さらに，括弧内を考慮または無視しても，文法的におかしくならない文となるように配慮すること．

7 金属の性質

Properties of Metals

Each metal or alloy has its own characteristic properties. Due to these properties, metals and alloys have been very important materials. The important properties of metals commonly assessed are as follows.

Tensile strength is the strength necessary to pull a metal testpiece into two pieces. This strength is obtained by dividing the maximum load by the original cross-sectional area.

Hardness is the resistance to being dented or penetrated. Main hardness testers for steel are the Brinell, Rockwell, and Vickers systems. Hardness of some steel can be greatly increased by heat treatment.

Ductility refers to the extent that the metal or alloy can be drawn out without breaking. The ductility of a material is given by elongation percentage or reduction percentage of area just after breaking. Aluminum, steel, and copper are very ductile materials.

Malleability refers to the extent that the metal or alloy can be hammered, rolled, or bent without cracking or breaking. The more malleable a material is, the easier it is hammered or rolled. Gold, silver, and tin have good malleability.

語　句

property[prɔ́pəti] 名性質　**metal**[métl] 名金属　**alloy**[ǽlɔi] 名合金　**due to**… のために＝owing to　**commonly**[kámənli] 副普通に← common 共通の　**assess**[əsés] 動評価する，課税する　**strength**[streŋθ] 名強度← strong 強い　**maximum** [mǽksiməm] 形最大限の←→ minimum 最小の　**hardness**[há:dnəs] 名硬度← hard 硬い　**resistance**[rizístəns] 名抵抗（性）　**dent**[dent] 名へこみ　動へこませる　**penetrate**[pénitreit] 名突っ込む，入り込む　**heat treatment** 熱処理　**ductility** [dʌktíliti] 名（金属の）延性　**elongation percentage** 伸び（率）　**malleable** [mǽliəbl] 形可鍛性の　**refer**[rifə́:] 動…のせいにする，…に帰する　**crack**[kræk] 名割れ，欠陥　動割る，砕く

●イントロ●　金属はほかの材料と違って金属光沢をもち，熱や電気をよく伝える．ここで述べる機械的性質も金属や合金の著しい特徴であって，金属をより有用なものにしている．Alloys もあわせて読まれたい．

構文 ▶**metals and alloys have been very important materials** ➡ have been は現在完了で継続の用法．「金属や合金は（引き続き）極めて重要な材料である」．

▶**Ductility refers to the extent that the metal or alloy can be drawn…** ➡この that は接続詞で同格の用法．「延性とは金属や合金が延伸されうるその程度についていわれることばである」．

▶**elongation percentage or reduction percentage of area…** ➡ elongation percent は伸び率（破壊時の伸びを元の長さで割ったもの）をパーセントで，reduction percentage は絞り率（破壊時の断面積の縮少を元の断面積で割ったもの）をパーセントで表したもの．共に簡単に伸び，絞りともいう．

▶**just after breaking** ➡「破壊した直後」．

▶**The more malleable…the easier～** ➡この「the 比較級＋ the 比較級」は関連している句であって，「より…であるときは，ますます～である」の意．「可鍛性に富むほど細工しやすくなる」．The more, the better.「多ければ多いほどよろしい」の意．

問 題

1. 次の（ ）の中の語を並べ変えて正しい英文とし，それを日本語に訳せ.

（1） Tensile strength refers to (a metal, can, breaking, the maximum stress, bear, before).

（2） Hardness refers to (the metal, to plastic deformation, that, can, the extent, resist).

（3） Ductility refers to (the metal, the extent, can, breaking, be drawn out, that, without).

（4） Malleability refers to (that, be bent, cracking, the extent, can, without, the metal).

2. 本文の内容と合っているものには○，合わないものには×をつけよ.

（1） Hardness is the most important property of metals.

（2） Heat treatment increases the hardness of some steels.

（3） Aluminum and copper have poor ductility.

（4） Gold and sliver are known to have good malleability.

コラム because と since の違い

理由や情報などを伝えるために，because や since（または as）の接続詞を用いて 2 文を繋げるが，これらの使い分けに注意する．聞き手・読者側が（理由や情報などの）内容を既に把握している際には，since や as を用いる．つまり，since/as に続く内容に重点はなく，結果側の文を主に伝えたいときに用いる．これとは対照的に，because で始まる節の（理由や情報などの）内容が聞き手・読者側にとって新しいものであり，これに重点を置いた文章となる．通常，since で始まる節は主節の前に，because で始まる節は主節の後に置く．

Since [As] a typhoon is approaching, the class is cancelled today.
（台風が近づいているため，本日は休講.）

I work hard on English, because my dream is to study abroad.
（英語を猛勉強しているのは，海外留学が私の夢だからである.）

8 合金

Alloys

An alloy is a metal mixture containing two or more metals, and sometimes contains nonmetals such as silicon, carbon, and others. Alloys are categorized into different phases: (1) solid solution, (2) an intermetalic compound, and (3) simple mixture of two or more metals, and combination of these phases are often found within the same alloy.

These days, almost all metallic materials are alloys. For instance, steel is an alloy of iron and carbon, brass is made of copper and zinc, and structural aluminum is made of aluminum, copper, magnesium, and others. By mixing metals we can obtain specific properties that are more attractive than those of pure metals. Some alloys have high strength; others have low melting points; others are used as refractory materials because of high melting point; and others have very strong resistance to corrosion.

These characteristics are brought about by changes in atomic-level structure, such as the penetration of atoms into other kind of atoms like a wedge. Generally speaking, an alloy is harder and has less electric conductivity than its constituent materials.

語　句

mixture[míkstʃə] 名 混合物　**contain**[kəntéin] 動 含む → container 容器　**nonmetal**[nɔnmétl] 名非金属　**silicon**[sílikən] 名けい素　**carbon**[ká:bən] 名炭素　**phase**[feiz] 名相，位相　**solid solution** 固溶体　**intermetallic**[intəmetɔ́:lik] 形金属間の　**combination**[kɔmbinéiʃən] 名結合← **combine** 結合する　**brass**[bræs] 名黄銅，真鍮　**structural**[strʌ́ktʃəral] 形構造上の　**aluminum**[əlú:mənəm] 名アルミニウム　**attractive**[ətræ̀ktiv] 形魅力のある　**melting point** 融点　**corrosion**[kəróuʒen] 名腐食　**bring about** 引き起こす　**atomic**[atɔ́mik] 形原子(力)の　**wedge**[wedʒ] 名くさび　**generally speaking** 一般的にいえば　**constituent**[kənstítjuənt] 名購造要素，成分

●イントロ● 金属は合金にすることによってその利用価値は著しく向上する．そのため金属は純金属としてだけでなく合金として広く利用されている．

構文 ▶**Alloys are categorized into different phases: (1) solid solution, (2)** …➡「合金は相の違いによって次のように分類される：(1) 固溶体，(2) …」．コロン（:）はこのように前述の内容を詳しく説明するときなどに用いられる．

▶**almost all metallic materials are alloys.** ➡「ほとんどすべての金属材料は合金である」．almost は副詞であるので直接名詞に前置されることはなく，このように almost all または almost every のようにする．

▶**specific properties that are more attractive than**…➡関係代名詞 that は which と互換的に使用されるが，先行詞に制限的形容詞（all, the…est, thesame など）がつけられているときに that がより好まれる．「…よりも魅力的な特別な性質」．

問 題

1. 次の（ ）の中の語を並べ変えて正しい英文とし，それを日本語に訳せ．
   (1) An alloy (containing, a metal mixture, two or more, is, metals).
   (2) An alloy (common, is, a pure metal, more, than).
   (3) An alloy (than, harder, its constituent metals, is).
   (4) An alloy (of, two or more metals, melting, and, is, mixing, made).
2. 本文の内容と合っているものには○，合わないものには×をつけよ．
   (1) An alloy always shows the same property as its constituent metals.
   (2) Steel is made of iron and carbon.
   (3) Brass is an alloy of copper and zinc.
   (4) Some alloys have very strong resistance to corrosion.

9 形状記憶合金

Shape-Memory Alloy

Shape-memory alloys have a very interesting and useful property in that, if deformed, they can recover their original shape upon heating. This strange behavior is called the shape-memory effect. When ordinary metal is strained beyond its elastic limit, permanent deformation results. Thus, if a bent wire straightens when being heated, it would be a quite unusual phenomenon. In the case of such alloys, however, the heating reminds the alloy of its original form. The effect results from the transformation of the martensite phase, therefore this alloy is sometimes called a marmen alloy. A typical marmen alloy is titanium and nickel（50：50）, nitinol alloy. It is presently used in small gadgets such as fastening pins, pipe connectors, eye-glass flames, cooler parts, brassieres, ash trays, and so on. However, shape-memory alloys have great potential for useful applications in the future. For instance, this material will be used in the fields of robotics, medical equipment, rockets, atomic power, assembly of integrated circuit, and so on.

語　句

memory[méməri] 名記憶，メモリー　**shape-memory alloy** 形状記憶合金　**permanent**[pə́:mənənt] 形永久的な　**martensite**[má:tenzait] 名マルテンサイト組織　**marmen alloy** マルメン合金（marmen は martensite memory の略）　**titanium**[taitéiniəm] 名チタン　**nitinol alloy** ニチノール合金　**gadget**[gǽdʒit] 名小型装置　**brassiere**[brəzíə] 名ブラジャー　**potential**[pəténʃəl] 名可能性

●イントロ●　形状記憶合金は普通の合金とは異なり，超弾性や形状記憶効果を示す特別な合金である．なお，超弾性とは外から力をかけて変形させても力を除去すると元の形に戻る性質である．また形状記憶効果とは外から力をかけて変形させてもある温度以上に加熱すると元の形に戻る性質である．

構文 ▶**in that, if deformed, they can recover…**➡ in that は接続詞句で「…という点で」. if defbrmed は仮定法.「その合金はもし変形しても，熱するともとの形状に回復するという点で」.

▶**if a bent wire straightens when being heated, it would be…**➡仮定法を受けて would を用いている． when being heated は分詞構文.「もしも曲った針金が熱することでまっすぐになるとしたら，それはまったく不思議な現象だ」.

▶**the heating reminds the alloy of…**➡ remind と of とは関連している.「熱することでその合金に元の形状を思い出させる」.

▶**…have great potential for useful applications**➡「…は将来有効に利用されるという点で大きな可能性をもっている」.

問 題

1. 次の（　）の中の語を並べ変えて正しい英文とし，それを日本語に訳せ.

（1） Shape-memory alloys (a very interesting, useful, and, have, property).

（2） Shape-memory alloys (convenience, us, very big, give).

（3） This effect (the martensite phase, from, the transformation, results, of).

（4） The bent wire (if, would, it, straighten, is, heated).

2. 本文の内容と合っているものには○，合わないものには×をつけよ.

（1） Shape-memory alloys have a unique property.

（2） Shape-memory alloys is made of iron and nickel.

（3） Shape-memory alloys is an alloy of titanium and zinc.

（4） Shape-memory alloys shows the same property as its constituent metals.

10 複合材料

Composite Materials

Composite materials are novel materials composed of two or more materials, each of which has its own superior property but at the same time lacks other properties. Each of the components should compensate for the inferior property of the combined material, and contribute to generating new and improved characteristics. New composite materials have found very wide market applications in industries such as for space, construction, and leisure.

Concrete reinforced with iron rods is a typical example of a composite material, in which iron rods have very high tensile strength while concrete can endure large compression forces in heavy structures. A new advanced composite material usually consists of two materials, one of which is a fiber and the other is the mother material. They are combined and then bonded by heat and pressure to achieve properties of high strength and stiffness. Other outstanding properties of such composite materials are their resistance to fatigue and corrosion. Superior fibers are carbon, aramid, and glass, and mother materials are usually thermosetting resins such as epoxies and polyimides.

語　句

composite[kɔ́mpəzit] 形合成の　名合成物　**novel**[nɔ́vl] 形新しい　**compose**[kəmpóuz] 動組み立てる(…of)　**compensate**[kɔ́mpənseit] 動補償する　**contribute**[kəntríbju:t] 動寄与する　**improve**[imprú:v] 動改善する　**space**[speis] 名空間，宇宙　**leisure**[léʒə] 名余暇　**concrete**[kɔ́nkri:t] 名コンクリート　**reinforce**[rì:infɔ́:s] 動強化する　**compression**[kəmpréʃon] 名圧縮　**advanced**[ədvá:nst] 形前進した　**mother material** 母材　**bond**[bɔnd] 名動接着(する)　**stiffness**[stífnis] 名強じんさ　**fatigue**[fatí:g] 名疲れ　**corrosion**[kəróuʒən] 名腐食　**thermosetting resin** 熱硬化性樹脂(→解説)

●イントロ●　最近の複合材料の発展は目覚ましい．複合材料としては炭素繊維で強化したプラスチック（FRP）が代表である．身近な例はテニスラケットやスキー板であろう．スキー板では温度，湿度，荷重，さらに滑走中の微妙な変化に対応できるように，何種類もの新素材が何層にも複合されてできている．

構文 ▶**materials composed of two or more materials** ➡「二つかそれ以上の材料から構成されている素材.」この composed of は過去分詞による形容詞の例.

▶**materials, each of which…** ➡関係代名詞 which の継続用法で先行詞は materials，したがって「材料のそれぞれは互いに補うようになっている.」複合材料は互いに弱いところを助け合っているので全体として強くなる.

▶**New composite materials have found very wide market application** ➡「新しい複合材料は非常に広範な応用分野を見出している」の意で，現在完了の継続の用法.

▶**iron rods have very high tensile strength while concrete can endure…** ➡ while「…しながら」，「…と同時に」の意.「コンクリートが大きな圧縮に耐えることができるのと同時に，鉄筋は非常に大きな引張り強さをもっている」.

解説 **新素材（Advanced material）**の多くは複合材料である．今までに数多く複合材料が開発され，今後もその用途は広がっていくことであろう．

熱硬化性樹脂（Thermosetting resin）であるフェノール樹脂，エポキシ樹脂，メラミン樹脂などは，熱を加えることによって化学結合が進行し，固化し，さらに加熱を続ければ分解してしまう．これに対して，ポリエチレンやポリ塩化ビニールなどのように熱を加えることによって軟化し時に溶けてしまうものを**熱可塑性樹脂（thermoplastic resin）**という．これらの分子の多くは繊維状であって熱を加えてもそれ以上化学反応は進行しない．

プラスチックが私達の生活に非常な便益を与えてくれているのは事実だが，同時に廃棄物公害の一因となっていることも見逃せない．

問 題

1. 次の（　）の中の語を並べ変えて正しい英文とし，それを日本語に訳せ.

（1） Composite materials (composed, usually, of, are, two or more materials).

（2） Composite materials (made, mixing and bonding, are, two or more materials, by).

（3） Composite materials (increase, to fatigue and corrosion, their resistance, can).

（4） A new advanced composite material (of, a mother material, a fiber, and, consists).

2. 本文の内容と合っているものには○，合わないものには×をつけよ.

（1） Components in composite materials do not usually compensate each other.

（2） Reinforced concrete is composed of concrete and iron rods.

（3） Iron rods in concrete have a role in increasing its compression strength.

（4） A resin is usually used for the mother material in composite materials.

11 炭素繊維

Carbon Fiber

Carbon fiber is a filament-shaped fiber with a diameter of about 6 to 10 micrometers. Fibers are classified into polyacrylonitrile（PAN）group, pitch group, and rayon group according to their original material.

The density of carbon fiber is usually as low as 1.8 g/cm^3, giving it a very large specific strength compared with other materials. It is also flexible, thermally and chemically stable, and is a good thermal and electrical conductor.

The principal use of high-performance carbon fiber is as a reinforcing component in composite materials. Carbon fiber in cement gives a highly superior construction material, that is, carbon fiber reinforced concrete（CFRC）.When it is mixed with epoxy resin, we get carbon fiber reinforced plastic（CFRP）.

Formerly, because of its high cost, carbon fiber was limitedly used in the aerospace industry in the United States, but recently it has come to be widely used in the sporting goods industry in tennis rackets, fishing rods, sailboats, skis, and so on.

語　句

filament-shaped [fíləmənt] 形フィラメント状の　**diameter** [daiǽmətər] 名直径　**fiber**[fáibər] 名繊維　**polyacrylonitrile** [pòlyacrylonítrile] 名ポリアクリロニトリル　**pitch**[pítʃ] 名ピッチ　**rayon**[réian] 名レイヨン　**flexible**[fléksəbl] 形柔軟な，曲げやすい　**principal**[prínsəpəl] 形主な　**epoxy resin**[ipáksi rézin] 名エポキシ樹脂

●イントロ●　炭素繊維（カーボンファイバ）は新素材として宇宙産業から身の回りの娯楽産業に至るまで広く利用されている．高強度，高弾力性，軽量でしかも熱にも薬品にも安定しているという優れた性質をもっている．特に炭素繊維は航空機の軽量化を支える複合材料として重要となってきている．

問題

1. 次の（ ）の中の語を並べ変えて正しい英文とし，それを日本語に訳せ．
   (1) Carbon fiber (used, as, is, a strengthening material).
   (2) Carbon fiber (extremely, consisting, is, a material, of, thin fibers).
   (3) FRP (is, composite material, famous, the most).
   (4) FRP (from tennis rackets, be, everywhere, can, to the aerospace industry, found).
2. 本文の内容と合っているものには○，合わないものには×をつけよ．
   (1) There are three types of carbon fibers.
   (2) Carbon fiber has a low specific strength compared with other materials.
   (3) FRP stands for a fiber reinforced plastic.
   (4) FRP is produced by mixing a carbon fiber with epoxy resin.

コラム 同意語（Synonym）

ほぼ同じ意味をもっている二つの語を互いに同意語という．例えば，generate と produce とは「生じる」の意の同意語である．しかし日本語で「生産」と「製造」の意味が異なるように，この二つの単語にはわずかなニュアンスの違いがある．generate はエネルギーや熱など非物質的なものに多く用いられ，produce は物質が生産するときに用いられる．velocity も speed と同意語であるが，velocity は単なる速さではなく，その方向まで考えた時に用いられる語で，日本語の「速度」に相当し，speed は単なる「速さ」である．日本語でもそうであるが，英語でも同意語を正確に使いわけることは容易ではない．

12 石油製品

Petroleum Products

Crude petroleum is a blackish, viscous liquid, and is a starting material for fuel and petrochemical materials. Crude petroleum is a mixture of various grades and types of hydrocarbons. It is separated by distillation into fractions such as gasoline, kerosene, heavy oil, and residues. Hydrocarbons from butane （C_4H_{10}） to C_{12} compounds appear in fuel for automobiles, and high boiling-point fractions C_{15} to C_{18} hydrocarbons are present in fuel for jet airplanes.

The process which transforms the crude fraction into gasoline using catalyst is called a catalytic reforming process. In addition to fuel for automobiles and jet airplanes, petroleum fuels can heat houses, and offices, and also be used to generate electric power. Residues can be used to yield familiar products as asphalt and coke.

語 句

crude[kru:d] 形天然のままの，粗雑の　**petroleum**[petróuliəm] 名石油　**viscous**[vískəs] 形粘る　**petrochemical**[pètroukémikəl] 形名石油化学の（製品）　**separate**[sépəreit] 動分離する　**distillation**[distiléiʃən] 名蒸留（法）← distill 蒸留する　**fraction**[frækʃən] 名成分，断片，分数　**gasoline**[gǽsəlì:n] 名ガソリン　**kerosene**[kérəsi:n] 名灯油（ケロシン）　**heavy oil** 重油　**butane**[bjú:tein] 名ブタン　**compound**[kɔ́mpaund] 名化合物　**process**[próuses] 名過程（プロセス），工程　**catalyst**[kǽtəlist] 名触媒　**reform**[rifɔ́:m] 動改善する，改良する　**factory**[fǽktəri] 名製造工場　**yield**[ji:ld] 動産出する，もたらす　**residue**[rézəd(j)ù:] 名残り，残留物　**familiar**[fəmíliə] 形よく知られている，普通の　**asphalt**[ǽsfælt] 名アスファルト　**coke**[kouk] 名コークス

●イントロ●　ガソリン，灯油，重油など私達の現在の生活は極めて密接な関係にある．それらは何からどのようにして作られるのか．

▶**It is separated by distillation into fractions** ➡「It（原油）は分留されてそれぞれの成分にわけられる」. separated into fractions となるべきところ途中に by distillation が挿入された形の文.

▶**high boiling-point fractions C_{15} to C_{18}** ➡「C_{15} から C_{18}（→解説）にいたる高沸点成分」. 普通は boiling point と分けてつづるがこの場合 fractions を形容していることを明瞭にするために間にハイフン（-）を入れてある.

▶**The process which transforms the crude fraction…** ➡ which は process を先行詞とする関係代名詞の限定用法で主格の働き.「原油成分をガソリンに変える操作」.

▶**In addition to fuel for…**➡「自動車やジェット機の燃料のほかに」.

解 説

C_{12}，C_{15}，など➡それぞれ炭素原子を 12 個，15 個もっている炭化水素の分子を意味している. ブタン C_4H_{10} のように簡単なものは水素原子の数を明示できるが，C が 12 個と多くなると二重結合があったりして，H の数がはっきりしないので，H の表示を略して C_{12} とだけ表すことがある. 飽和していれば $C_{12}H_{26}$ と書ける.

原油の主成分は C と H だけであるが，そのほか S，N，O などが含まれる. なかでも，S は燃やしたときに SO_2（二酸化硫黄）となり大気汚染の原因となる. 最近では S 除去の技術が開発され，日本の空も大変きれいになった. 原油から除去された S は回収硫黄といわれ，硫黄の資源として用いられる.

改質操作（Reforming process）➡重油よりもガソリンのほうが需要が多く値段が高い. そのため，C_{18} のように長い炭素の鎖を断ち切って，2 個の C_9 に変化させれば有利である. 水素を通じ触媒を使って行うこのような分解操作を，改質操作という. 水素を通じない分解操作を接触分解（cracking：クラッキング）という. ともに良質のガソリンを得ることができる.

問題

1. 次の（　）の中の語を並べ変えて正しい英文とし，それを日本語に訳せ．
   (1) Crude petroleum (of hydrocarbons, a mixture, of various grades, is).
   (2) It (separated, various fractions, by distillation, is, into).
   (3) Hydrocarbons (in fuel, to C_{12} compounds, appear, from butane).
   (4) Residues (used, asphalt and coke, can, be, as, to yield, familiar products).
2. 本文の内容と合っているものには○，合わないものには×をつけよ．
   (1) Crude petroleum is called a gasoline.
   (2) Petroleum fuels are used to generate electric power.
   (3) A fuel for cars is produced from a crude oil by a catalytic reforming process.
   (4) Hydrocarbons from C_{15} to C_{18} compounds appear in fuel for airplanes.

コラム　一つの文に一つのこと

技術英文では，1文で一つのことを伝えるように心掛けるとよい．慣れている日本語では，複数のことを1文に盛り込むことは容易かもしれないが，技術英文には適さない．最も シンプルで読みやすい文章構成は，「SVO. SVO. SVO. …」で各主語（S）は直前のSまたはOに関連している形であろう．

○● 章末問題 ●○

1. 空欄を埋めて，日本語に訳せ.

(1)　The worker must have information (　　　) its precise shape and dimensions.

(2)　Correct lines are used (　　　) proper places.

(3)　This instrument must be handled (　　　) carefully (　　　) (　　　).

(4)　They can be categozied roughly (　　　) two types.

(5)　A limit gauge is a double gauge (　　　) "go" size and "no go" side.

(6)　The stress-strain diagram is one way (　　　) show the relation (　　　) stress (　　　) strain.

(7)　This strength is obtained (　　　) dividing the maximum load (　　　) the original cross-sectional area.

(8)　Steel is an alloy (　　　) iron and carbon and brass is made (　　　) copper and zinc.

(9)　Shape-memory effect results (　　　) the transformation of the martensite phase.

(10)　Concrete reinforced (　　　) iron rods is a typical example.

(11)　The density of carbon fiber is usually (　　　) low (　　　) 1.8 g/cm^3.

(12)　Crude petroleum is the starting material (　　　) petrochemical materials.

2. 次の文を英語に直せ.

(1)　少なくとも，二つの図面が使われる.

(2)　寸法の数字は寸法線の上の真ん中に書かれる.

(3)　マイクロメータのスピンドルは正確に機械加工されたねじである.

(4)　ブロックはセットで売られている.

(5)　限界ゲージは寸法測定において早くて正確な方法を与える.

(6)　外力や応力に対する材料の忍耐力を知ることはとても重要だ.

(7)　引張強さは金属試験片を二つに引き裂くために必要な強さである.

(8)　合金は二つまたはそれ以上の金属を含む金属の混合物である.

(9)　この奇妙な現象は形状記憶効果と呼ばれる．

(10)　複合材料は二つまたはそれ以上の材料から構成される新しい材料である．

(11)　高性能な炭素繊維の主な使われ方は複合材料における強化成分である．

(12)　原油は炭化水素のさまざまなグレードの混合物である．

III

機械工作

Machining

この章で述べる内容は，従来の機械工学または機械技術といわれるものの中核をなす部分である．以前は，機械の教育といえば工作機械の使用法に習熟することがその中心であった．

私達の身近に見られる品物で，それがどのような方法で加工されているのかわからない場合が少なくない．たとえばアルミサッシはどのように造られたのか，また，台所にあるなべやかまの加工法は何か．それらは工業人だけではなく一般の人達にとっても常識とすることが望ましい．

1 機械

Machine

A machine is defined as a combination of rigid bodies forced to make definite motions and being capable of performing useful work. Since the work referred to here is equal to energy, a machine must always receive energy from the input side and deliver work in a useful form to the output side. According to the definition, a camera, which has very fine mechanisms and does useful jobs, is not a machine for it never transforms energy through it.

Though computers or word processors really have neither any movable parts nor transmit successful mechanical work as their final output, these days they are considered to be machines. They can convert information in the form of rough and vague input into very elegant and useful package for us. For instance, if we input some data into a computer, we can immediately obtain more useful answers as output. Therefore, today, the definition of machine is formulated as a kind of a device which converts energy or information from a less desirable condition to a more desirable one.

語　句

define[difáin] 動定義する　**rigid**[rídʒid] 形堅い，固定した　**force**[fɔ:s] 名力　動強制する　**perform**[pəfɔ́:m] 動遂行する→performance パフォーマンス　**input**[ínput] 名入力　**deliver**[dilívə] 動引き渡す　**output**[áutput] 名出力　**mechanism**[mékənizm] 名機械装置，メカニズム　**successful**[səksésfəl] 形成功した　**elegant**[éligənt] 形上品な，洗練された　**data**[déitə] 名データ（datum の複数形だが単数扱い）　**immediately**[imí:djətli] 副直ちに　**answer**[á:nsə] 名動答え（る）

●イントロ●　エネルギーの転換機という古典的な機械の定義が拡張され，現在ではエネルギー以外に情報をも取り扱うようになった事情を述べている．

▶**being capable of performing useful work.** ➡「有用な仕事を遂行することができる」．capable of は便利な表現で，a room capable of seating 100 students「100 人の生徒を収容できる教室」のように使える.

▶**Though computers or word processors really have neither…nor〜**➡ neither と nor とは関連して用いられ,「…もなければ〜もない」の意.「コンピュータやワードプロセッサは可動部分をもっていないし，また機械的仕事を遂行しないが」.

▶**from less desirable condition to a more desirable one.**➡ from と to とは関連しており，less desirable と more desirable とも関連している.「より望ましくない状態からより望ましい状態へ」.

問題

1. 次の（　）の中の語を並べ変えて正しい英文とし，それを日本語に訳せ.
   (1) A machine (forced, specific motions, has, rigid bodies, to make).
   (2) A machine (capable, performing, is, useful work, of).
   (3) A machine (deliver work, must, energy, and, receive, in a useful form).
   (4) A machine (converts, a device, is, that, defined, as, energy or information).
2. 本文の内容と合っているものには○，合わないものには×をつけよ.
   (1) A camera is a precise machine with fine mechanisms.
   (2) A car is a wellknown machine that converts energy into mehcnaical energy.
   (3) A computer isn't a machine because it doesn't have any movable parts.
   (4) A computer is capable of processing information to produce a desired output.

2 旋盤

Lathe

A lathe is a typical machine tool. A machine tool is defined as a machine for cutting metals. Usually it must firmly keep a cutting tool and steadily rotate a workpiece.

We are living in the machine age, surrounded by jet airplanes, steamships, locomotives, computers, and automobiles to give us a very convenience. All of these devices were made possibly through the development of the lathe. The lathe is the oldest and the most important machine tool of all.

A workpiece is securely gripped within the lathe by a holding device called a chuck, and it rotates under power against a suitably fixed cutting tool. Removed shavings are called chips, which can be dangerous because of their sharp edges. Workpieces in various forms such as a cone, sphere, concentric body, as well as a true cylinder, can be held by various types of chucks. Machining operations performed by the lathe include facing, boring, turning, and threading.

語句

lathe[leið] 名旋盤　**cutting**[kʌ́ting] 名切断，切削←cut 切る　**steadily**[stédili] 副確実に　**rotate**[routéit] 動回転する→rotation 回転　**workpiece**[wə́:kpi:s] 名工作(物)　**surround**[səráund] 動取り囲む　**steamship**[stí:mʃip] 名汽船　**locomotive**[lóukəmətiv] 名機関車　**convenience**[kənví:njəns] 名便利　**development**[divéləpmənt] 名進歩，開発←develop 進歩する　**grip**[grip] 動名(しっかり)つかむ(こと)，柄　**chuck**[tʃʌk] 名(旋盤の)チャック　**suitably**[sú:təbli] 副適当に　**chip**[tʃíp] 名(旋盤の)チップ，切りくず　**sharp**[ʃa:p] 形鋭い，とがった　**cone**[koun] 名円すい体　**sphere**[sfiə] 名球(体)　**concentric**[kənséntrik] 形中心を共有する　**facing**[féisiŋ] 名面削り　**boring**[bɔ́:riŋ] 名穴あけ　**turning**[tə́:niŋ] 名外丸削り　**threading**[θrédiŋ] 名ねじ切り

●イントロ● 工作機械（machine tool）の中で最も一般的な旋盤について述べている．旋盤はコンピュータ制御の導入によりNC旋盤またはCNC旋盤として広く使用されている．Ⅳ章のNC工作機械についてもあわせて読まれたい．

構文 ▶**We are living in the machine age, surrounded by…** ➡ surrounded by…は分詞構文で，because we are surrounded by…と書き換えられる．「私達は機械の時代に生きている．それは私達がジェット機，機関車，…によって囲まれていることでもわかる」．

▶**a holding device called a chuck,** ➡「チャックと呼ばれる保持用工具」．過去分詞形calledによってdeviceを修飾している．

▶**it rotates under power against a suitably fixed cutting tool.** ➡主語はit，動詞はrotatesであるから「S＋V」の構造．under以下は副詞句．「それは正しく固定された切削工具に向き合うよう，動力によって回転する」．

▶**chips, which can be dangerous because of their sharp edges.** ➡ whichは継続用法．「切りくずは鋭い刃先を持っているので危険である」．because ofの後ろにはこのように句が来る．becauseを使ってbecause they have sharp edges. としてもよい．

▶**as well as a true cylinder** ➡ as well asはandの強調用法．

解説 金材材料の加工は大きく二つにわけられる．金属の塑性（plasticity）（金属をそのまま，または熱を加えて変形すること）と，金属の被削性（cuttability）とを利用する場合がある．鋳造，鍛造，プレス加工は前者で，一般に精度はよくないが，大量生産に適する．これに対して旋盤，フライス盤，研削盤による加工は後者の例で，材料の不必要な部分を刃物（cutting tool）で削り取る加工（切削加工）である．加工品の精度はよいが作業能率が悪く，材料のむだが多い．工作機械は，機械をつくる機械といわれ，上記の材料の被削性を応用する．工作機械の中でも旋盤は最も利用度の高い機械である．

問 題

1. 次の（ ）の中の語を並べ変えて正しい英文とし，それを日本語に訳せ．
   （1） A lathe（a machine tool, a piece of, used, is, to shape, metal）.
   （2） A lathe（the, important, is, machine tool, most）.
   （3） A workpiece（shaved, by, a material, is, the lathe）.
   （4） A workpiece（must, before cutting, securely, held, be, by a chuck）.
2. 本文の内容と合っているものには○，合わないものには×をつけよ．
   （1） A lathe is a modern machine tool.
   （2） A cone-shaped workpiece cannot be made by a lathe.
   （3） We are surrounded by a lot of machines.
   （4） Many machines are made possible through the development of the lathe.

コラム 一つの技術文章には一つの主題

技術英文による文章を書く場面は，技術報告・マニュアル（説明書）・提案書・仕様明細書などのほか，学術論文や学位論文などさまざまである．いずれにも共通していえることは，書くべき内容と目的が必ず存在することである．まさに“徒然なるままに”書く技術文章など，ほぼありえない．学術論文や学位論文では，その自由度の高さから，内容が散漫になってしまい，主題（読み手に伝えたいメッセージ）が不明確になりがちである．論文作成前には，書きたい内容に近い先行研究や関連文献を調べ，よく読み解き，自らの立ち位置と主題を明確にしてから，執筆を行うことが肝要である．

3 フライス盤

Milling Machine

A milling machine is an important machine tool. In some way the milling machine is like a grinding machine. It has a cutting tool, called a milling cutter, which is basically a wheel with many cutting edges. Thus, it is often called a multiple point cutting tool. In contrast to the lathe, the milling cutter rotates while its workpiece is fixed on the table. When the work to be machined is fed into the cutter, chips of metal are cut away.

Milling machines are usually classified as follows：(1) horizontal milling machine；(2) horizontal universal milling machine, which is the same as the horizontal type except that its table swivels about its vertical axis；and (3) vertical milling machine.

Milling machines provide a wide-range of operations for cutting such as flat, curved, or irregular surfaces, slots, grooves, keyways, cams, or many other shapes. Very complex curved surfaces, such as flutes on a drill or teeth on gears, are cut by means of a spiral milling cutter.

Fig. 17　Milling cutter

語　句

milling[míliŋ] 名切削，製粉　**grinding machine** 名研削盤　**basically**[béisikəli] 副基本的には←base 土台　**cutting edge** 切れ刃　**thus**[ðʌs] 副このようにして　**contrast**[kɔ́ntræst] 名対照　**classify**[klǽsifai] 動分類する→classification 分類　**horizontal**[hɔrizɔ́ntl] 形水平な←horizon 水平線　**universal**[ju:nivə́:səl] 形万能の←universe 宇宙　**except**[iksépt] 前…を除いては，…のほか　**swivel**[swívl] 名回転継ぎ手　動回転する　**vertical**[və́:tikl] 形垂直の　**axis**[ǽksis] 名軸　**provide**[prəváid] 動供給する　**wide-range** 広範囲　**irregular**[irégjulə] 形不規則な⟷regular 規則的な　**slot**[slɔt] 名細長い穴，スロット　**groove**[gru:v] 名みぞ，決まった習慣　**keyway**[kí:wei] 名キー溝　**cam**[kæm] 名カム　**flute**[flu:t] 名(ドリルの)縦溝，フルート　**teeth**[ti:θ] 名 tooth(歯)の複数　**gear**[giə] 名歯車　**spiral**[spáirəl] 形らせん状の

●イントロ●　英語では milling machine といい，フライスとはいわない点に注意する．フライス盤の語源はフランス語の小型フライス盤，fraise にある．

▶**In some way the milling machine is**… ➡「ある意味ではフライス盤は研削盤に似ている」の意.

▶**a multiple point cutting tool** ➡旋盤に使うバイトは刃先が一点であるのに対し，フライス盤の刃物はたくさんの切れ刃をもっている.「多点回転刃物」ということがある．points となっていないのは multiple point が cutting tool の形容詞として使われているため.

解説　最も利用度の高い工作機械は旋盤だと前述したが，次に多く利用されているのがフライス盤である．旋盤とフライス盤の基本的な違いは，前者が「工作物回転，刃物固定」であるのに対し，後者は「工作物固定，刃物回転」である点である．しかし運動は相対的なものであるから切削の理論は同一である.

問　題

1. 次の（　）の中の語を並べ変えて正しい英文とし，それを日本語に訳せ.
   (1) A milling machine (a multiple point cutting tool, uses, usually).
   (2) Milling machines (classified, three, are, different types, into).
   (3) Milling machines (for cutting, a wide-range, provide, of operations).
   (4) A complex curved surface (prepared, a milling machine, can, by, be).
2. 本文の内容と合っているものには○，合わないものには×をつけよ.
   (1) A milling machine is very similar to a lathe.
   (2) In some ways a milling machine is like a slot machine.
   (3) A milling machine is the most important of all machine tools.
   (4) A workpiece is fixed and a milling cutter rotates in a milling machine.

4 ボール盤

Drill Press

Every machine shop, holes of many sizes must be made in metal parts. Some of these holes must be very smooth and straight with an exact size and accurate location. Others may not need precise size and location. In any way, holes must be made with proper tools such as a drilling machine or drill press. It is important to obtain the knowledge of and skill with a drilling machine and tools for the practice in a machine shop.

Although many other cutting tools are used, the twist drill is the most prominent. Whether the drill is large or small, carbon or high-speed steel, held in a chuck or otherwise, and used with a jig or without, the principles of its operation are nearly the same.

If many pieces of the same kind have to be drilled, a drill jig can save a lot of time and money. It is a tool for holding the workpiece while it is being drilled, and at the same time it can guide a drill to the exact position specified. Of course, considerable fund may be needed to prepare a jig, but the savings in time and effort would pay for the money expended.

語　句

location[loukéiʃən] 名位置　**prominent**[prámənənt] 形突出した，卓越した　**high-speed steel** 高速度鋼（ハイス）　**drill jig** ドリルジグ　**fund**[fʌnd] 名資金

●イントロ●　日本ではボール盤といっているもので，小さい手回しのドリルからラジアルボール盤といわれる大きいものまで使われている．いずれも穴あけの単能機械である．

構文 ▶**Whether the drill is large or small, carbon or high-speed steel, held in a chuck or otherwise, and used with a jig or without, the principles…** ➡長い文だが，the drill is large…から with a jig

or without までは全部 whether にかかっている.「ドリルが大きいか小さいか，炭素鋼か高速度鋼か，チャックによってかまたは別の方法で保持されているか，さらにジグを使うか使わないか，にかかわらずその原理は…である」.

問題

1. 次の（　）の中の語を並べ変えて正しい英文とし，それを日本語に訳せ.
   (1) A drilling machine (called, a drill press, is).
   (2) A drill press (in metal, is, to make, a machine, holes).
   (3) A twist drill (with, to remove, is, two twisted flutes, shavings, a drill).
   (4) A drill jig (holding, a special tool, is, for, the workpiece).
2. 本文の内容と合っているものには○，合わないものには×をつけよ.
   (1) A drill press is so easy that the operator doesn't need any special knowledge and skill.
   (2) The drill jig is very expensive.
   (3) A drill jig is usually only for the beginner.
   (4) A drill jig is normally used for efficient drilling operations.

コラム　一つのパラグラフには一つの話題

綿密に計画立てられたパラグラフ（段落）構成は，技術文章において重要であり，各パラグラフ内での論理構成も大事な要素である．文章全体では一つの主題（→ p.76）があり，その主題の導出・裏付け・考察などをするために各役割のパラグラフが文章中で構成されることになる．各パラグラフでは役割に沿った話題を一つ提供し，無暗に複数の話題を盛り込まないように心掛けて執筆するとよい．パラグラフは適度な長さ（5～10 文程度）にすると読みやすい．また，パラグラフ内の論理構成が曖昧であると，文法的に正しくても読み手に“英語が下手”という印象を与えかねないので粗末にしないこと．

5 研削盤

Grinding Machine

Grinding produces very exact and fine surfaces. There are two types of grinding machines; that is, for a flat surface and for a cylindrical surface. Wheels with very fine grains of abrasive can be used to produce a workpiece to extremely close tolerances. The grinding process is often used on metals too hard to machine otherwise.

Flat surface grinders are classified into a horizontal spindle type and a vertical spindle type. The most common type is the horizontal spindle grinder. The work is mounted securely on the table either in a vise, or with a magnetic chuck. The grinding wheel is mounted on a horizontal spindle which may be raised or lowered at intervals of one-thousandth of one millimeter. The depth of cut is determined by the position of the grinding wheel.

The lengthwise movement of a table is called work speed. The cross travel of a table is called feed. For rough grinding, fast work speed and coarse feed are adopted. For finish grinding, slow work speed and fine feed are usually employed.

語　句

exact[igzǽkt] 形正確な　**fine**[fáin] 形美しい，細かい　**cylindrical**[silíndrikəl] 形円筒状の　**wheel**[hwi:l] 名車輪，ホイール　**grain**[grein] 名粒子　**abrasive**[əbréisiv] 名研磨材　**tolerance**[tɔ́lərəns] 名許容差　**otherwise**[ʌ́ðəwaiz] 副違った方法で　**mount**[maunt] 動据えつける　**securely**[sikjúəli] 副安全に　**vise**[vais] 名万力（＝vice）　**magnetic**[mægnétik] 形磁石の　**raise**[reiz] 動立たせる，持ち上げる　**lower**[lóuə] 動低くする　**interval**[íntəvl] 名距離，休止期　**determine**[ditə́:min] 動決定する　**position**[pəzíʃən] 名位置　**lengthwise**[léŋθwaiz] 形副長い（く）　**feed**[fi:d] 動供給する，食物を与える　名送り，フィード　**coarse**[kɔ:s] 形粗い，ざらざらの ⟷ smooth　**adopt**[ədɔ́pt] 動採用する　**employ**[implɔ́i] 動雇用する，使用する

●イントロ●　回転砥石による研削は，砥粒が細かいことと高速回転による研削とで極めて精密な仕上げが期待できる．ハイテク機器の部品の精密仕上げには現在のところ研削盤の活躍によるところが大きい．

▶**on metals too hard to machine otherwise.** ➡「too…to～」の構文で，「あまり硬くてほかの方法では加工できないような金属に用いられる」．

▶**at intervals of one-thousandth of one millimeter.** ➡「1 mm の 1/1000 の精度で上下することができる」．

▶**lengthwise movement** ➡「長手（横）方向への運動」．帯状のものがあるときに長手方向を lengthwise または lengthways という．それと直角の方向は widthwise か widthways を使う．

解説

許容差（Tolerance）　許容差（公差ともいう）の考え方は現在の大量生産にとって極めて重要である．自動車を構成する何万という部品がピタリと組み立てられるためには，それぞれが図面の寸法どおりに仕上げられていなければならない．しかし機械の精度にばらつきがあり，工員の技術にも差があるのでそれはほとんど不可能である．そこで許容差の概念が生まれる．

図面の寸法の上下に許される幅を作ってその寸法内の部品であれば合格とする．合格した部品なら，そのまま手直しをせずに組み立てられる．そうして部品の互換性が保証される．

目標寸法 25 mm 直径の丸棒の許容寸法が 25 ± 0.05 mm であるとき，± 0.05 mm を許容差という．これを目標寸法の百分率で表して $\pm 0.05 \div 25 \times 100 = \pm 0.2$ %とすることもある．

問 題

1. 次の（ ）の中の語を並べ変えて正しい英文とし，それを日本語に訳せ.
   （1） A grinding machine (for short, called, a grinder, is).
   （2） A grinding machine (is, an abrasive wheel, with, a machine).
   （3） There (of, two, grinding machines, types, are).
   （4） A magnetic chuck (used, to hold, is, the tool, the workpiece).
2. 本文の内容と合っているものには○，合わないものには×をつけよ.
   （1） A grinding machine can only produce a flat surface.
   （2） A horizontal spindle grinder is the most common flat surface grinder.
   （3） The work speed is the speed of the rotating grinding wheel.
   （4） Slow work speed and fine feed are necessary to obtain a smooth surface.

コラム the の使いどころ

定冠詞の the は，直後の名詞で示すものが一つに決まる場合に用いられる．しかし，名詞や序数で決まるのではなく，前後の文章や文意によっては不定冠詞や不定複数が適切な場合もある．

the world（この世界）	a world of future（未来の世界）
the only solution（唯一解）	an only child（一人っ子）
the third meeting（第三回会議）	a third party（第三者）

発音は通常「ザ」であり，不定冠詞 a/an と同様に母音発音が続く場合と，特に the（唯一無二であることや，「あの○○」と決まること）を強調したい場合は「ジ」と発音する．

6 金属切削の機構

Basic Mechanism of Metal Cutting

The basic types of chip-type machining processes are shown, in their simplest form, in Fig. 18. The detail of cutting form is shown in Fig. 19. The tool, which has a certain rake angle α and relief angle, moves along the surface of the workpiece at a depth of t_1. The material in front of the tool is continuously sheared along the shear plane, which makes an angle of ϕ with the surface of the workpiece. This shear angle and the rake angle contribute combinedly to determine the chip thickness, t_2. The ratio of t_1 to t_2 is called the cutting ratio r. The relationship among the shear angle, the rake angle, and the cutting ratio is given by the following equation.

$$\tan\phi = r\cos\alpha/(1 - r\sin\alpha)$$

It can be easily seen that the shear angle is important because it controls the thickness of the chip. Generally speaking, the larger the shear angle, the less the deformation of the chip and the smoother the cutting operation.

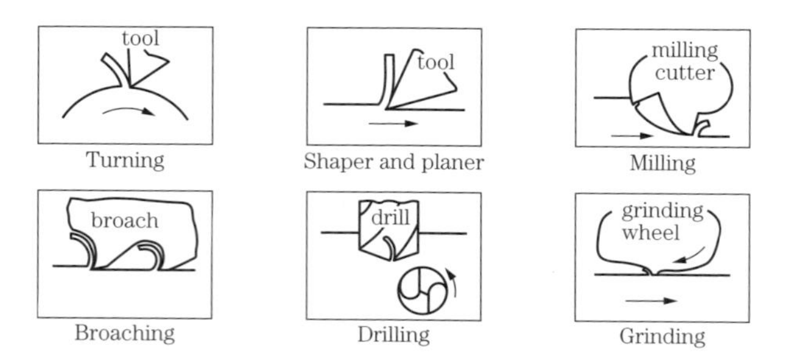

Fig. 18　Chip-type machining process

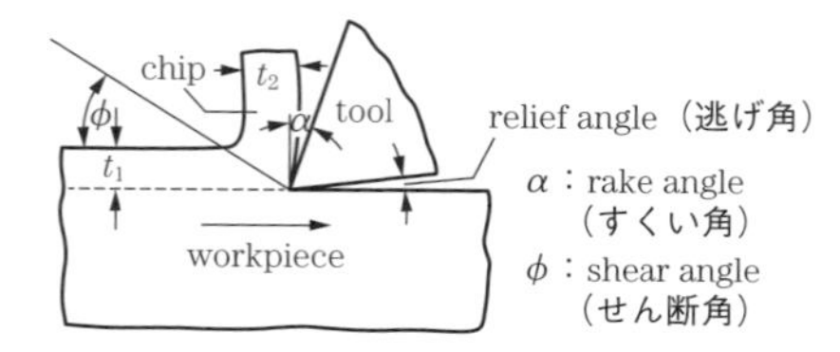

Fig. 19　Basic mechanism of cutting process

語　句

basic[béisik] 形基礎の ← base 土台　**rake angle** すくい角　**relief angle** 逃げ角　**depth**[depθ] 名深さ ← deep 深い　**in front of** …の前に　**continuous**[kəntínjuəs] 形連続的な　**shear**[ʃiə] 動はさみ切る　**shear angle** せん断角　**contribute**[kəntríbju:t] 動貢献する　**ratio**[réiʃiou] 名比，比率　**relationship**[riléiʃənʃip] 名関係　**equation**[ikwéiʃən] 名方程式 ← equal 等しい　**smooth**[smu:ð] 名なめらかな

●イントロ●　切削はさまざまな工作機械で行われるがその理論はほぼ共通している．理論によってチップの形状や必要動力などが推定できる．

構文

▶**The tool, which has a…relief angle, moves** ➡ which から angle までを（　）に入れて考える．「すくい角 α と逃げ角を持った刃先が表面に沿って動く」．この which は限定用法と考える．

▶**The ratio of t_1 to t_2** ➡「t_2 に対する t_1 の比」．式では t_1/t_2 となる．

▶**The relationship among the shear angle, the rake angle, and thecutting ratio** ➡「せん断角，すくい角と切削比との間の関係」．このように among は 3 者以上の関係についていい，2 者の間のときは between を使う．

▶**It can be easily seen that…** ➡ that 以下の節の内容を形式主語 It が受けている．

▶**the larger the shear angle, the less the deformation…and the smoother～** ➡形容詞の比較級が三つ連続しているがそれぞれ関連している．「せん断角が小さいほど，…の変形が少なくかつ～が円滑にすすむ」の意.

問　題

1. 次の（　）の中の語を並べ変えて正しい英文とし，それを日本語に訳せ．
   (1)　There (the shear angle, is, and, the rake angle, among, the cutting ratio, a relation,).
   (2)　The tool (moves, the surface, of the workpiece, to remove, along the surface).

（3）　The material（sheared, in, the tool, is, front, of, continuously, to make chips）.

（4）　The shear angle（control, the thickness, can, the chip, of）.

2. 本文の内容と合っているものには○，合わないものには×をつけよ．

（1）　The ratio of t_1 to t_2 is expressed by the formula t_2/t_1.

（2）　The basic mechanism of cutting is only applicable to the cutting of a lathe.

（3）　The larger the shear angle, the more the deformation of the chip.

（4）　The larger the shear angle, the smoother the cutting operation.

コラム　as a result の逆が because

「結果として，…」を表すとき，日本人の陥りやすい誤訳として安易に as a result を使いがちである．原因と結果を示す 2 文間を繋げるために as a result が用いられるべきである．もし（原因と結果の）2 文の順番が逆になったとき because が相応しくなる場合でないと，as a result は適当でない．

誤用：The specimen was tested. As a result, our prototype satisfies the criteria.（試験片の検査をした．その結果，試作品は基準を満たした．）

※日本語訳に違和感はないが，日本人にしか通じないかもしれない．

前後の文を入れ替えて，As a result の代わりに Because を入れると

Our prototype satisfies the criteria. Because the specimen was tested.（試作品が基準を満たした．なぜなら試験片の検査をしたからだ．）

となり，不自然な内容となる．これは 2 文が原因と結果の関係に無いからである．

正しい使い方の例を以下に記す．

Fiber-reinforced glass was used. As a result, our prototype satisfies the criteria.（繊維強化ガラスを用いた．その結果，試作品は基準を満たせた．）

⇔ Our prototype satisfies the criteria. Because, fiber-reinforced glass was used.（試作品が基準を満たした．なぜなら，繊維強化ガラスを用いたからである．）

7 潤滑剤

Lubricant

A lubricant is liquid or solid which is used to prevent direct contact of parts in relative motion, and it can reduce friction and wear. At the same time, the lubricant plays a role in cooling parts, and the rust prevention is also important.

Various types of lubricants are prepared from crude petroleum by fractional distillation. Superior lubricant oils are synthetically manufactured, such as ester-type or silicone oil-type. Solid-type lubricants such as graphite and talc can be used widely in machinery. In case of solid lubricants, they usually have a microstructure composed of laminar type molecules.

Lubricants mainly work by creating a considerably thick oil layer between two surfaces. If the layer is too thin, seizure of the moving parts will occur. It must also withstand the load imposed by a heavy part, so a suitable viscosity of the lubricant must be selected in accordance with the weight of parts being in relative motion.

語句

lubricant[lú:brikənt] 名潤滑油(剤) **prevent**[privént] 動妨げる，防ぐ **relative** [rélətiv] 形相対的な **relative motion** 相対運動 **reduce**[ridjú:s] 動減少する，引き下げる **friction**[fríkʃən] 名摩擦 **wear**[wεə] 名動摩耗(する) **role**[roul] 名役割 **superior**[sju:píəriə] 形優秀な⟷ inferior 劣った **synthetically**[sinθétikəli] 副合成的に **manufacture**[mænjufǽktʃə] 名動製造(する) **silicone**[sílikoun] 名シリコーン(けい素を主成分とする高分子重合体．シリコン(けい素)silicon と混同しないこと) **graphite**[grǽfait] 名黒鉛，グラファイト **talc**[tælk] 名滑石，タルク **machinery**[məʃí:nəri] 名(集合的に)機械(類) **microstructure** 微細構造 **laminar**[lǽminə] 形薄片状の **molecule**[mɔ́likju:l] 名分子 **layer**[léiə] 名層 **seizure**[sí:ʒə] 名(機械の)焼付き **withstand**[wiðstǽnd] 動もちこたえる **impose** [impóuz] 動課する，押しつける **viscosity**[viskɔ́siti] 名粘度 **in accordance with** …に一致して，に従って ＝according to

●イントロ● 潤滑剤は「縁の下の力持ち」的な存在であるが，各種機械には焼付け防止のために絶対に欠くことのできないものである．

構文

▶**liquid or solid which is used to prevent direct…** ➡ which は liquid or solid を先行詞とする主格関係代名詞の限定用法．「…を防ぐのに用いられる液体，または固体である」．

▶**parts in relative motion** ➡「相対運動をしている部品」．

▶**…are prepared from crude oil by fractional distillation.** ➡「…は分留操作によって原油から得られる」．

▶**a microstructure composed of…** ➡「平面型の分子から構成されている微細構造」．microstructure の反対は macrostructure「巨視的（マクロ）構造」．

▶**parts being in relative motion** ➡上記の parts in relative motion と似ているが being があると「現在相対運動をしている」の意になる．

問 題

1. 次の（ ）の中の語を並べ変えて正しい英文とし，それを日本語に訳せ．
   - (1) A lubricant (to prevent, is, direct contact, used, of two parts, in relative motion).
   - (2) A lubricant (reduce, friction, and, wear, can, of the parts).
   - (3) A lubricant (can, rusting, prevent, from, the parts).
   - (4) A lubricant (creating, between, works, by, a thick oil layer, two parts).
2. 本文の内容と合っているものには○，合わないものには×をつけよ．
   - (1) Synthetically manufactured lubricants are inferior to natural ones.
   - (2) Ester-type and silicone-type lubricants are solid-type lubricants.
   - (3) The thickness of lubricants must be suitable according to the load between parts.
   - (4) When heavy parts are handled, high viscosity lubricants are needed.

8 溶接

Welding

Pieces of metal may be fastened or joined together with mechanical fasteners such as nuts and bolts. Another way to join them is to weld them together. One kind of weld is made by heating the edges of the pieces to be joined. The metal at the edges melts and blends or fuses together. In this case, metal is usually added to the joint from a filler rod which has also been melted. This rod is usually made of the same material as the welded pieces. The pieces of metal to be joined are called the base metal. The heat may be provided by burning gases (as in the case of oxyacetylene welding), or with electric current (arc welding).

Forge welding or pressure welding is another way to join metal pieces. The area to be welded are heated until the metal pieces reach a plastic state, but are not fused. Then the pieces are forced together by pressure (usually by hammering).

Resistance welding uses the heat generated by electric current passing through a small area of metal being joined. This type of welding is sometimes called spot welding because the site of the weld appears as a spot, not a seam.

語句

fasten[fάːsn] 動固着させる→ fastener ファスナ　**weld**[weld] 動溶接する　**blend** [blend] 動混ぜ合わせる ＝mix　**fuse**[fjuːz] 動融解する ＝melt　**filler rod** 溶加棒　**base metal** 母材　**oxyacetylene welding** 酸素アセチレン焔溶接　**arc welding** アーク溶接　**forge welding** 鍛接　**pressure welding** 圧接　**plastic state** 塑性状態　**hammer**[hǽmə] 動(ハンマで) 打つ　**resistance** [rizístəns] 名抵抗 ← resist 抵抗する　**spot welding** スポット溶接　**site**[sait] 名場所，敷地　**seam**[siːm] 名縫い目 動縫う

●イントロ●　溶接は造船，建築，機械などだけではなく，最近は美術工芸関係にも利用されるようになってきた.

▶**another way to join them** ➡すぐ前の文で「ボルトナットによる接合を述べた後なので another といっている．「それらを接合する別の方法」．

▶**from a filler rod which has also been melted** ➡ which の先行詞は filler rod．現在完了形を使っているのは，溶加棒が既に融解された状態になっているため．「融解された溶加棒から」．

▶**the same material as…** ➡ the same と as とは関連している．「溶接される部材と同じ材質の」．

▶**are heated until the metal pieces reach…** ➡「金属片が塑性状態に達するまで熱せられる」．until「…まで」と by「…までに」との違いに注意する．I will be back by 5 O'clock this evening.「夕方 5 時までには戻ります」．

解説

二つのものを接合することは難しい．紙や木片などは接着剤で接着することができる．この場合の接着剤は被接着物とは異質の物質なので，どうしても接着力が弱いという欠点がある．ところが溶接は，被接着物を熱で溶かし，そのすき間に同じ材質の溶加棒を溶かして流し込んで固めるため，極めて強固な接着ができる．現在は溶接の技術が進み，船舶のようなかなり厚い部材まで溶接できるようになった．またコンピュータと組んでロボットによる溶接も非常に進んでいる．

溶接に似ているものに**ろうづけ（brazing）**がある．これは接着する金属の間に融解した異質の金属を流し込んで接着するものであり，ハンダろうによるろうづけが代表である．接着力は弱いが，手軽で安価なのでエレクトロニクス関係をはじめ各方面で行われている．

問題

1. 次の（　）の中の語を並べ変えて正しい英文とし，それを日本語に訳せ．
   (1) Welding (often, done, melting, is, metal pieces, by, to join them, together).
   (2) The pieces of metal (that, will, are, be welded, the base metal, called).
   (3) Forge welding (another, metal pieces, to join, is, way).
   (4) Forge welding (fuse, the metal pieces, that, will, does not, be joined).
2. 本文の内容と合っているものには○，合わないものには×をつけよ．
   (1) Pieces of metal can be fastened by either mechanical fasteners or welding.
   (2) A filter rod does not need to be the same as the base metal.
   (3) Resistance welding uses the heat generated by burning gases.
   (4) We call this resistance welding spot welding.

コラム　よく使われる動詞 provide

「～を与える」「～を提供する」などの意味合いをもつ provide は，技術英語でよく使われる動詞である．一般的な動詞 give と似ているが，give は受取り側の存在が意識されている（第5文型 SVOO の）印象がある．一方，provide は受取り側を意識せずに SVO の文型を取ることが多く，技術英語で好まれる簡潔さをもつ．日本語で「～を提供する」と訳すと，違和感があることが多いので，注意．

The data provide a reasonable estimate.（データから妥当な推定ができる．）

Provide this field with your PIN.（この欄に暗証番号を入力せよ．）

9 鍛造

Forging

Forging refers to the plastic deformation of metal, usually at high temperature, into the desired shape by compressive force. The simplest form of forging is flattening the metal, which has been heated, by compressing between two flat parallel platens. This operation is called upsetting and is carried out by a hammer, anvil, swage block and others. For large scale forging, a hydraulic press is used to provide sufficient force.

Like the fiber structure of wood, metal has anisotropic structure when it is forged with very strong pressing force. This means that forged metal has greater tensile strength and tougher property than that shaped by cutting operations. Tools such as spanners or nippers, wheels for locomotives, and connecting rods in automobile engines are all forged to give them great strength and toughness. In practice, forgeability is infuenced by strength, toughness, ductility, and other properties of the material to be treated. Ranking various metals according to their forgeability gives us the order of aluminum alloy, magnesium alloy, copper alloy and carbon steel.

語　句

forge[fɔ:dʒ] 動鍛造する（仕事場）　**deformation**[difɔ:méiʃən] 名変形← deform 変形する　**simple**[símpl] 形簡単な　**flatten**[flǽtn] 動平らにする← flat 平らな　**platen** [plǽtn] 名圧盤プラテン　**upset**[ʌpsét] 動熱した鉄材をつぶす，アプセットする　**swage** [sweidʒ] 名スエジ　**hydraulic press** 油圧機　**anisotropic**[ənisətrɔ́:pik] 形異方性の←→ isotropic 等方性の（→解説）　**spanner**[spǽnə] 名スパナ　**nipper**[nípə] 名ニッパ　**connecting rod** 名連接棒，コンロッド　**rank**[ræŋk] 動等級をつける　**forgeability**[fɔ:dʒəbíliti] 名鍛造性

●イントロ●　鍛造でつくられた製品は，一般に強度が優れている．その理由は，材質中の微細な結晶粒子が鍛造されている間に，引き伸ばされて繊維組織を作るためであると考えられている．

▶Forging refers to…, ➡ refer to は be よりもやや婉曲で，「…ということができる」程度の表現．「鍛造は塑性状態における金属の変形ということができる．」

▶Like the fiber structure of wood ➡「木材の繊維構造と似て」．

▶This means that forged metal has… ➡ that は接続詞で名詞節を導くもの．「このことは，鍛造された金属は大きい引張強さを有するということ（名詞節）を意味している」の意．

▶In practice ➡「実際の鍛造では」の意．

▶Ranking various metals according to their forgeability… ➡ Ranking は動名詞で主語の働き．「可鍛性に着目していろいろな金属を順序づけすると…」．

解説 **異方性（Anisotropy）と等方性（Isotropy）**　iso（等しい）＋tropic（変化）の合成語で，an- は否定を表す．等方性とは物質のどの方向に対しても同じ性質（たとえば光の伝達速度）を示すような物質をいい，異方性とは測定する方向によって異なる測定値を与えるような物質をいう．一般に結晶は異方性を示すが，ガラスなど結晶が認められないものは等方性である．金属は微細な結晶からできているが，結晶の向きがばらばらであるために異方性を示さない．ところが，鍛造しているうちに結晶がある方向に引き伸ばされたり，切断されたりして，一方向に繊維状に並び異方性を示すようになる．その結果，鍛造品は強度が増すと考えられている．

急冷などの方法により意図的に結晶の成長を妨げて作った等方性物質を**アモルファス物質（無定形物質）（amorphous substance）**という．優れた性質をもつアモルファス物質（半導体など）が作られており，将来が注目される素材の一つである．

問　題

1. 次の（　）の中の語を並べ変えて正しい英文とし，それを日本語に訳せ．
   （1）　Forging (metals, to, the plastic deformation, refers, of).
   （2）　Forging is a process that (heating, shapes, by, metal, and, hammering, the metal).
   （3）　A forged metal (tensile strength, has, greater, and, toughness).
   （4）　Tools such as spanners and nippers (forging, usually, produced, are, by).
2. 本文の内容と合っているものには○，合わないものには×をつけよ．
   （1）　We need only a hammer for forging metals.
   （2）　A forged metal has an anisotropic structure like wood.
   （3）　Forgeability depends on the strength of the material forged.
   （4）　Carbon steel is the most forgeable metal.

コラム　句動詞表現よりも好まれる 1 語動詞

句動詞（phrasal verb）とは「動詞＋副詞」または「動詞＋（副詞）＋前置詞」の 2〜3 語で構成され，まとまって一つの動詞的な役割をする言い回しである．比較的に一般的な動詞と副詞・前置詞で構成され，インフォーマルまたは口語的な印象がある．これとは対照的に，1 語動詞（single verb）ではフォーマルな印象を与え，また語数が少なくなることからも，特別な事情のない限り，技術英文では限定的な 1 語動詞での表現が好まれる．以下に，句動詞から 1 語動詞に言い換えた例を示す．

「爆発する」blow up ⇒ explode　　「〜を減らす」cut down on ⇒ reduce
「〜を実行する」carry out ⇒ execute　「〜を調べる」look into ⇒ examine

10 金属の鋳造

Metal Casting

Metal casting means to pour molten metal into a cavity or a mold to form it into a desired shape. Upon cooling, the metal solidifies and takes the shape of the cavity or mold. Objects made by this method are called castings. When the casting is taken from the mold, cleaning and finishing are needed because its surface is usually rough.

Casting offers several advantages over other metal forming, including adaptability to a complex shape, large objects, and mass production. Castings are not stronger than products made by forging, because the casting is thought to have less fiber structure than forgings.

The mold means a cavity into which molten metal is poured. The mold is usually made from sand or heat-resisting metal. Sand molds have the advantage of a low cost, but they can only be used once. In contrast, metal molds can be used repeatedly and give a smooth surface of castings.

語　句

pour[pɔ́ː] 動注ぐ　**cavity**[kǽvəti] 名穴　**mold**[mould] 名型，鋳型　**solidify**[səlídəfai] 動凝固する　**casting**[kǽstiŋ] 名鋳物，鋳造　**adaptability**[ədæptəbíliti] 名適合性，順応性

●イントロ●　金属の鋳造とは，金属やプラスチックを高温で溶かし，型へ流し込んだり圧入したりして成形する加工法である．

▶**including adaptability to a complex shapem, large objects, and mass production.** ➡ including 以下は分詞構文．「これらの利点には，複雑な形，大きな物体，および大量生産への順応性が含まれている」．

▶**from sand or a heat-resisting metal.** ➡ from は heat-resisting metal へもかかる．「砂あるいは耐熱性の金属からつくられる」の意．

III 機械工作

多くの寺にあるつり鐘や仏像は青銅の鋳物でつくられている．鋳型に溶けた金属を流し込み，ある形の製品をつくる技術は昔からあった．日本で奈良時代に，既に鋳造の技術が確立されていたことは驚くべきことである．刀鍛冶や鉄砲鍛冶などの日本古来の優れた技術が，現在の繁栄と密接に関係していると考えられる．

問　題

1. 次の（　）の中の語を並べ変えて正しい英文とし，それを日本語に訳せ．
   (1) Casting is a process that (pouring, by, creates, molten metal, something, into, a mold).
   (2) The mold for casting (made, sand or metal, is, usually, from).
   (3) Casting (over, advantages, offers, several, other metal forming process).
   (4) Castings (are made, objects, which, are, casting, by).
2. 本文の内容と合っているものには○，合わないものには×をつけよ．
   (1) The surface of the castings is always smooth.
   (2) Sand molds have the advantage of a low cost.
   (3) Metal molds can be used only once.
   (4) Castings are stronger than products made by forging.

コラム　さまざまな否定の語句

否定の表現にはさまざまな形がある．品詞や用法に注意して使い分けよう．

not 副詞	never 副詞	hardly 副詞
no 副詞・形容詞	neither 副詞・形容詞	no more 形容詞
none 代名詞	nothing 代名詞	nobody 代名詞
unless 接続詞	without 前置詞	nowhere 名詞・副詞
few 形容詞（可算名詞）	little 形容詞（不可算名詞）	

11 冷間圧延鋼材

Cold-Rolled Steel

Cold-rolled steel is made from hot-rolled steel which has been rolled while still hot. Hot-rolled steel is slightly larger than the final desired size of the steel product. The thickest product of hot-rolling is called the bloom, and through hot- and cold-rolling operation, the thickness of the bloom is reduced to a billet, a slab, a plate, a sheet, and finally a foil.

For cold rolling, hot-rolled steel is immersed in water containing sulfuric acid, in a pickling process, in order to remove the black skim or scales from the surface of the steel. The sulfuric acid is washed off by dipping the steel slab or other shaped steel in pure water and then in lime water. When dried, the objects are rolled while cold between highly polished rollers under great pressure. This gives them a smooth bright finish and a very exact size. These products are called cold-rolled sheet and are often used without any further finishing or machining.

語　句

roll[roul] 動圧延する　**hot-rolled** 形熱間圧延した　**slightly**[sláitli] 副少し，わずかに　**volume**[vɔ́ljum] 名体積，かさ　**bloom**[blu:m] 名角柱鋼材（大型），ブルーム　**billet**[bílit] 名角柱鋼材（中型），ビレット　**slab**[slæb] 名厚鋼板，スラブ　**plate**[pleit] 名鋼板　**sheet**[ʃi:t] 名鋼板　**foil**[fɔil] 名極薄鋼板　**immerse**[imə́:s] 動浸す，沈める　**pickling**[píkliŋ] 名浸漬　**sulfuric acid** 硫酸　**skim**[skim] 名薄皮　**scale**[skeil] 名酸化物の皮膜　**dip**[dip] 動浸す　**lime water** 石灰水　**polish**[pɔ́liʃ] 動みがく　**bright**[brait] 動輝く，光る　**further**[fə́:ðə] 副より遠い，さらに

●イントロ●　自動車や家電製品などに大量に消費される薄鋼板は圧延によって造られる．圧延とは鋼を2本のローラーの間にはさんで圧力をかけ，だんだんと薄くしていく操作で，熱間の圧延と冷間の圧延とがある．

▶**steel which has been rolled while still hot.** ➡「まだ熱いうちに圧延された鋼」．which は steel を先行詞にした限定用法．

▶**through hot- and cold-rolling operation** ➡ hot- は hot-rolling の略された形．「熱間および冷間の操作によってブルームは次第に薄板にされる」．

▶**in order to remove the black skim or…** ➡「黒皮や…を取り除くために」．in order to は単に to でもほとんど同意だがやや強調的に用いられる．

▶**…and then in lime water.** ➡「まず硫酸溶液を洗い落し，それから石灰水に浸す」の意．

▶**highly polished rollers** ➡ highly は rollers ではなく polished を修飾するので本来ならば highly-polished となるべきところだが，highly が -ly 副詞であるので（-）をつけなくてもよい．hot-rolling operation と比較せよ．

解説

溶鉱炉（高炉）（Blast furnace）　溶鉱炉から得られる鉄鋼は炭素を多量に含んでいる銑鉄であり鋳造はできるが圧延はできない．それを転炉に入れ，炭素を燃やして炭素分を少なくしたものを鋼（C が 0.7～2 %）という．鋼は圧延が可能になり，また熱処理で非常に優れた性質を与えることができるので，私達の身近な鉄はほとんど鋼である．

問題

1. 次の（ ）の中の語を並べ変えて正しい英文とし，それを日本語に訳せ．
   - (1) The bloom (a mass, to be rolled, is, of, iron or steel).
   - (2) Rolling is a process where (rolls, the bloom,passed, is, through).
   - (3) Rolling (by, classified, into, two types, is, the rolling temperature).
   - (4) Cold-rolled steel (made, is, from, hot-rolled steel).
2. 本文の内容と合っているものには○，合わないものには×をつけよ．
   - (1) Cold-rolled steel is thinner than hot-rolled steel.
   - (2) Hot-rolled steel has a smooth bright surface.
   - (3) Cold-rolled steel is immersed in an acid to removing the scales.
   - (4) Cold-rolled steel can be used without any further finishing.

12 熱処理

Heat Treatment

Heat treatment is an operation or combination of operations, involving heating and cooling of metal or alloy without melting, for the purpose of giving certain desirable properties to the metal or alloy. The procedures of heat treatment are classified as hardening, tempering, annealing, and case hardening.

Hardening means making material harder. Carbon steel is hardened by slowly heating to a near cherry red color, and then suddenly cooling in air, oil, water, or other liquid.

Tempering means the re-heating of steel after the hardening operation, and then cooling gradually to remove the brittleness and to give it the desirable hardness and toughness.

Annealing means to make hardened steel softer and to remove its brittleness. It is the opposite of hardening. In hardening, heated steel is cooled as quickly as possible, but in annealing it is cooled as slowly as possible.

Case hardening is performed by penetrating carbon into surface of low carbon steel to give hardness and wear-resistant properties to the surface only. Gears and other parts are often case-hardened.

語句

heat treatment 熱処理　**desirable**[dizáiərəbl] 形望ましい　**hardening**[há:dniŋ] 名焼き入れ ＝quenching　**tempering**[témpəriŋ] 名焼戻し　**annealing**[əní:liŋ] 名焼鈍し　**case hardening**[kéis-hɑ:dniŋ] 名表面硬化　**cherry red** 桜色の　**suddenly**[sʌ́dnli] 副突然に　**gradually**[grǽdjuəli] 副次第に，徐々に　**brittleness**[brítlnis] 名脆性　**toughness**[tʌ́fnis] 名じん性　**opposite**[ɔ́pəzit] 名反対（の物）　**quickly**[kwíkli] 副素早く　**possible**[pɔ́sibl] 形可能な，ありうる　**wear-resistant** 耐摩耗性の

●イントロ●　材料をさまざまな条件の下で加熱あるいは冷却して，材料の性質を変えることを熱処理という．現在ではいろいろの材料に対して行われる．

▶**operations, involving heating and cooling of metal…** ➡ involving 以下は分詞構文で，operations の内容を説明している．「金属または合金を融解せずに加熱または冷却するような操作」．

▶**for the purpose of…** ➡「…の目的で」．

▶**…cooling gradually to remove the brittleness and to give it…** ➡「脆性を取り除き，望ましい強度とじん性を与えるために…」．

▶**by penetrating carbon into the surface…** ➡「…の表面に炭素を浸入させることによって」．この操作を浸炭（carburizing）といい，表面だけが硬くなる．

解説 熱処理をすると材料が硬くなるのは原子・分子・微結晶などが特別な並び方をするためである．電車の中の乗客は電車が急発進したり急停車したりする度に揺れ動くが，もし乗客がお互いにスクラムを組んで足を踏ん張っていればある程度耐えられよう．熱処理で硬化するのもそれに似ている．原子や分子が互いに緊張した状態にあるので外力に耐えることができる．

問　題

1. 次の（　）の中の語を並べ変えて正しい英文とし，それを日本語に訳せ．
   （1） Heat treatment (heating, and, cooling, involves, of metal).
   （2） Heat treatment (alter, the metal, can, of, the properties).
   （3） Carbon steel (suited, for, is, particularly, heat treatment).
   （4） Tempering (removing brittleness, is, to, annealing, in regard to, similar).
2. 本文の内容と合っているものには○，合わないものには×をつけよ．
   （1） Hardening means to make the steel harder by quenching.
   （2） Annealing is the same procedure as hardening.
   （3） Annealing means to make hardened steel harder.
   （4） Case hardening is a process where the surface of the metal is hardened.

○● 章末問題 ●○

1. 空欄を埋めて，日本語に訳せ.

(1) A machine must receive energy and deliver work (　　　) a useful form.

(2) A machine tool is defined (　　　) a machine for cutting metals.

(3) A milling machine is like grinding machine (　　　) some ways.

(4) Holes must be made (　　　) proper tools such as a drilling machine or dirll press.

(5) There are two types (　　　) griding machines.

(6) The shear angle contributes (　　　) determine the chip thickness.

(7) A lubricant is liquid or solid which is used (　　　) prevent direct contact of parts.

(8) Resistance welding uses the heat generated (　　　) electric current.

(9) Forging refers (　　　) the plastic deformation of metal.

(10) Casting offers several advantages (　　　) other metal forming.

(11) Cold-rolled steel is made (　　　) hot-rolled steel.

(12) Hardening means (　　　) make material harder.

2. 次の文を英語に直せ.

(1) カメラは，それを通してエネルギーが全然変化しないから機械ではない.

(2) 旋盤は最も古くそして最も重要な工作機械である.

(3) フライス盤は，工作物がテーブルに固定され，刃物が回転する.

(4) その操作の原理はほとんど同じです.

(5) 研削盤はあまりに硬くてほかの方法では加工できない金属によく使用される.

(6) 工具の前にある材料が連続的にせん断面に沿ってせん断される.

(7) 潤滑剤は二つの表面の間にかなり厚い油の層を作ることによって働く.

(8) 溶加棒はふつう溶接される部品を同じ材料で作られる.

(9) 金属は強い圧縮力で鍛造されたとき異方性の組織をもつ.

(10) この方法でつくられた物体は鋳物と呼ばれる.

(11) 熱間圧延の厚い製品はブルームと呼ばれる.

(12) 焼き戻しは焼き入れ操作の後で鉄鋼を再加熱することを意味する.

IV

機械工学の現在

Advanced Mechanical Engineering

ほとんどの新しい技術がそうであるように，最近の機械工学はコンピュータを抜きにしては語れない．コンピュータやセンサを駆使した機械工学の新しい技術について述べよう．

ここで次の点にご留意いただきたい．現在開発中のものは，その評価が必ずしも確立していない．あるものはますます脚光を浴びるかもしれないし，また，あるものはそれほど発展することもなく消えてしまうかもしれない．その意味で読者諸君がその消長にしっかりと注目していただきたい．

1 キャド（コンピュータ支援設計）

CAD

CAD is an acronym for "computer-aided design," and refers to the use of computers for designing or planning the optimum shape and performance of products. CAD has a vast range of application from tiny integrated circuits (IC) to large maps covering huge continents. Newly developed CAD systems contain functions allowing not only for designing but also creating models necessary for technical estimation. Compared with conventional manual designing and drawing, such new CAD system have the following advantages : (1) easy modification of drawing and redrawing ; (2) partial reuse of drawing data ; (3) correct and rapid arrangement of complicated drawings ; and (4) easy enlargement and transformation of drawings.

Thanks to making good use of CAD system, the productivity of drawing work has been remarkably improved. Formerly, almost every CAD system had to be operated by a large-scaled computer, but recently CAD system can run on smaller 32-bits or 64-bits computers, so-called personal computers. Thus, in the near future drawing boards and instruments in a drawing room may be replaced by a personal computer with sophisticated CAD softwares.

語　句

acronym[ǽkrənim] 名頭字語　**aid**[eid] 動援助する　**optimum**[ɔ́ptiməm] 名最適条件　形最適の　**performance**[pəfɔ́:məns] 名性能　**integrate**[íntigrèit] 動統合する　**integrated circuit** 集積回路　**continent**[kɔ́ntinent] 名大陸　**model**[mɔ́dl] 名模型，モデル　**estimation**[estiméiʃən] 名評価← estimate 評価する　**redraw**[ridrɔ́:] 動再び製図する　**partial**[pá:ʃəl] 形部分的な　**complicated**[kɔ́mplikeitid] 形複雑な　**enlargement**[inlá:dʒmənt] 名拡大← enlarge 大きくする　**transformation**[trænsfəméiʃən] 名変化← transform 形を変える　**remarkably**[rimá:kəbli] 副目立って　**so-called** いわゆる　**drawing board** 製図板　**replace**[ripléis] 動…に取って代わる　**sophisticated**[səfístikeitid] 形巧妙な，洗練された

●イントロ●　現在どこの工場でもCADの基本を知らずに製品設計を語ることはできないであろう．最近，CADはCAMと一体になりつつある．

構文

▶**CAD is…and refers to the use of～**➡「CADは…であり，さらに～の利用に関するものである」．

▶**a vast range of application from…to～**➡ fromとtoとは関連している．「…から～に至る広い応用の分野」．

▶**Compared with…**➡ beingが省略された分詞構文である．「従来の手作業の設計や製図に比較すれば」．

▶**Thanks to making good use of…**➡ thanks to…はdue toやowing toと同じく，「…のおかげで，…のせいで」の意．「CADシステムを有効に活用すれば」．

▶**drawing boards and instruments may be replaced by…**➡「製図板や製図機械がパソコンによって取って代わられる」．

問題

1. 次の（　）の中の語を並べ変えて正しい英文とし，それを日本語に訳せ．
　(1)　CAD (an acronym, for, is, "computer-aided design").
　(2)　CAD (refers, for, the use, to, designing or planning systems, of computers).
　(3)　The productivity (has, by, improved, been, using CAD, remarkably).
　(4)　Drawing boards and instruments (replaced, by, may, be, a personal computer).
2. 本文の内容と合っているものには○，合わないものには×をつけよ．
　(1)　CAD can be used only for designing small integrated circuits.
　(2)　CAD allows not only for designing but also for creating models.
　(3)　Conventional manual designing has many advantages.
　(4)　Nowadays, CAD can be used for your office or home.

2 メカトロニクス

Mechatronics

The word, mechatronics, was coined in Japan, and is so-called Japanese English, but it enjoys wide use in the United States and Europe now. It can be defined as a new form of enginnering in which mechanical engineering and electronic engineering are closely combined to give various new functions to machines. New machines require new level of high precision, high speed, and high dexterity. These functions had previously been provided by a physical mechanism such as a link mechanism or a cam system. These days, however, such conventional mechanisms have been superseded by electronic devices such as sensors, microcomputers, and actuators.

Typical examples equipped with these advanced electronics are robots, machining centers, VTR, and so on. System engineering aims at the development of large-scaled systems usually relying upon a super computer, while the modern mechatronics usually involves a smaller microcomputer installed on each machine and its versatility mainly depends on software technology rather than hardware.

語　句

mechatronics[mækətrɔ́niks] 名メカトロニクス　**coin**[kɔ́in] 動作り出す　**Japanese English** 和製英語　**function**[fʌ́ŋkʃən] 名機能　**require**[rikwáiə] 動要求する　**dexterity**[dekstérəti] 名巧妙さ　**cam system** カム機構　**supersede**[su:pəsí:d] 動…に取って代わる　**sensor**[sénsə] 名感知器，センサ　**microcomputer**[máikròkompu:tə] 名マイクロコンピュータ　**actuator**[ǽktjueitə] 動アクチュエータ　**equipped with** …を備えた　**robot**[róubɔt] 名ロボット　**machining center** マシニングセンタ　**VTR** ビデオテープレコーダ　**install**[instɔ́:l] 動備え付ける　**versatility**[və:sətíliti] 名多才多能　**software technology** ソフトウェア技術　**hardware**[hɑ́:dwɛə] 名ハードウェア，ハード

●イントロ●　メカトロニクスは日本でつくられた英語だが，現在では世界各国で用いられている．今後ますます発展すると思われる分野である．

構文 ▶**The word, mechatronics, was coined in Japan, and is so-called Japanese English,** ➡「メカトロニクスということばは日本でつくられた，いわゆる和製英語である.」The word と mechatronics とは同格.

▶**it enjoys wide use in…** ➡「それは…で幅広い使用を楽しんでいる」とは「…で広く用いられている」の意.

▶**engineering in which…** ➡ which の先行詞は直前の engineering.「その技術の中で機械工学と電子工学とが密接に結びついている」.

▶**…have been superseded by electronic devices such as～** ➡「…は～のような電子機器装置によって置き換えられている.」継続の現在完了で受動態の用法.

▶**software technology rather than hardware.** ➡ rather than は「…よりはむしろ～.」したがって「ハードウェアよりはむしろソフトウェア技術に依存している」の意.

解説 **ソフトウェアとハードウェア（software and hardware）**　ハードウェアは古いことばで，Webster の辞典によると 1515 年に使われ始め，“ware made of metal”「金属製の品物」である．これに対して，ソフトウェアは 1962 年に使われ始めたことばで，“the entire set of programs, procedures, and related documentation associated with a system”.「システムに関連するプログラム，操作，および関係文献の集成」である．かつてはコンピュータの本体をハードウェアと呼び，その使用法に関する各種文献をソフトウェアと呼んでいたが，現在はコンピュータの分野を越えて日常の生活にまで入り込んできている．「この機械買ったんだけどソフトがよくわからない」などと使う．

問 題

1. 次の（　）の中の語を並べ変えて正しい英文とし，それを日本語に訳せ．

（1） The motion in a machine (transformed, a physical mechanism, was, through).

（2） Mechatronics (a combined technology, of, is, mechanics and electronics).

（3） Robots (for, the typical examples, are, of, one, mechatronics).

（4） Electronic devices (played, in, very important, have, roles, mechatronics).

2. 本文の内容と合っているものには○，合わないものには×をつけよ．

（1） Mechatronics usually rely on a super computer.

（2） The word mechatronics was coined in the United States of America.

（3） A link mechanism or a cam system has been improved by mechatronics.

（4） Modern mechatronics has put an importance on hardware technology.

コラム 冗長な語句を除去した 1 単語表現

凝った英文を作ろうとして，冗長表現（redundant expression）を招くことがある．たとえば，「腹痛が痛い」のように，（程度の差こそあれ）腹“痛”は痛いのであるから，後半の「が痛い」は不要かもしれない．しばしば見かける下記の英語表現のように，括弧内の語句は除去可能であり，そのほうが語数も少なくて読みやすい．

(uniformly) homogeneous　　spherical (in shape)
(final) conclusion　　few (in number)
(in the month of) April　　red (in color)

3 センサ

Sensor

When carrying out a cutting operation on a machine tool, we have to decide the depth or feed of a cutter by means of observing the graduation scaled on a handle. With the automatization of machine tools, devices which can judge the position or moving speed of a cutting tool or object in replacement of human senses are essential. These devices not only detect such physically measured value, but also can change such values into another signal and transmit it into other devices. They also sometimes carry out compensating and eliminating of errors, or numerical operations. These devices are called sensors.

A device called a limit switch, a sort of sensor, is often used in an electrical appliance. A sensor can detect the contact of one object with another as shown in Fig. 20. Using the sensor, we can open or close the electric circuit, by which the movements of an object such as stopping, starting, accelerating, or decelerating are easily controlled. The most commonly adopted sensor is a visual sensor in a camera, which allows automatically judgement of distance and fine focus of target objects.

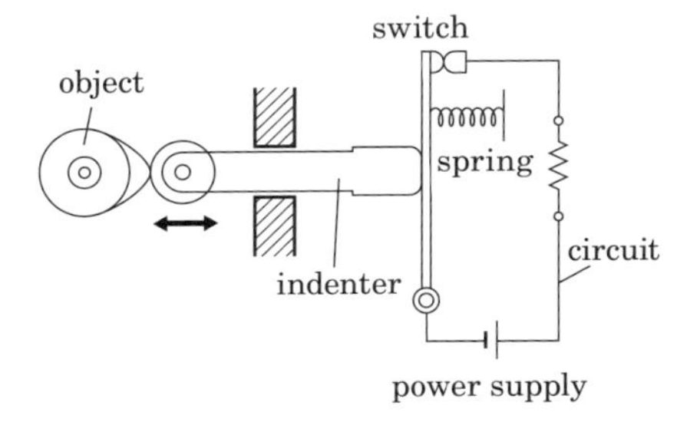

Fig. 20　limit switch

語　句

carry out 実行する　**observe**[əbsə́:v] 動観察する　**scale**[skeil] 名動目盛り（をつける）　**automatization**[ɔ:təmətaizéiʃən] 名オートメーション化　**replacement** [ripléismənt] 名交換，交替← replace 交換する　**essential**[isénʃəl] 形必要な　**physically**[fízikəli] 副物理的に　**signal**[sígnəl] 名合図，シグナル← sign 合図　**limit switch** リミットスイッチ　**detect**[ditékt] 動探知する　**contact**[kɔ́ntækt] 名動接触（する）　**visual**[vízjuəl] 名視覚による　**judgement**[dʒʌ́dʒmənt] 名判断←judge 判断する　**focus**[fóukəs] 名焦点　**target**[tá:git] 名動目標（となる）

IV 機械工学の現在

●イントロ●　検出器では何のことかわからないほどセンサはすっかり日本語になっている．センサのあらましを学ぶ．

構文

▶**When carrying out cutting operation…, we have to～** ➡ when は接続詞．carrying out…，は分詞構文．「工作機械で切削作業をするときには」．

▶**With the automatization of machine tools, …essential.** ➡長い文だが基本的には devices are essential で「S＋V＋C」の構造．「工作機械の自動化に伴って，…の装置が必要になる」の意．

▶**These devices not only detect…but also can change～** ➡ not only と but also とは関連している．「これらの装置は物理測定量を検出するだけではなくそれらの値を～に変換することができる」．

▶**a sort of sensor** ➡「一種のセンサ．」a kind of でも同意．

▶**Using the sensor, we can…** ➡分詞構文で，「このセンサを利用すれば我々は…できる」と訳す．

▶**the electric circuit, by which the momements of an object such as…** ➡ which の先行詞は electric circuit．such as は for example と同じく例示する時に使う．「その電気回路のおかげで，工作物の…のような動作が容易に制御される」．

問　題

1. 次の（　）の中の語を並べ変えて正しい英文とし，それを日本語に訳せ．
   (1) The devices that (be, instead, used, of, can, human senses) are called sensors.
   (2) A sensor (and, it, receives, in a distinctive manner, responds, a signal).
   (3) A thermocouple (converts, is, temperature, to, a sensor, which, voltage).
   (4) A limit switch (used, a sensor, is, electrical appliances, in).
2. 本文の内容と合っているものには○，合わないものには×をつけよ．
   (1) Sensors are especially important in the field of mechatronics.
   (2) A camera has a visual sensor which judges the distance.
   (3) A limit switch is a sensor which can detect the contact of one object with another.
   (4) The automatization of machine tools requires a sensor to judge the position.

コラム most と most of の違い

「～のほとんどは」「～の多くは」と言い表すために most または most of (the) が用いられる．of を挟むかどうかで使う場面が異なることに注意する．対象のすべてを指したり，一般的なことを述べたりするときは形容詞 most で（対象の）名詞を修飾する．特定のグループ（the が付く）に限定する場合は most of（the）の形をとる．

Most people desire to live long.
((一般的に）人々のほとんどは長生きしたいと願っている．)
Most of the people in this country live longer than 70 years.
（この国に住む人々の多くは 70 年以上の長生きをする．）

4 フィードバック制御システム

Feedback Control System

An open-loop control system utilizes a controller in order to obtain the desired value (Fig. 21). In contrast, a closed-loop control system utilizes the deviation obtained by comparing actual output value with input desired value in order to control the system (Fig. 22). The route from the sensor point, where the output value is measured, to the controller, where the deviation is added to a manipulatable target, is calld a feedback route. This route, which can only transmit information, is the key part of all devices involving a feedback control system.

In the case of steering an automobile, the driver uses his or her sight to visually catch and compare the actual position of the car in relation to his or her destination. In this case, the driver performs the role of a sensor, controller, and actuator for steering wheel. Familiar household electrical appliances have such a basic closed-loop configuration. For example, a refrigerator is equipped with a temperature setting knob to input desired temperature, a thermostat to measure the actual temperature, comparing device, and controller.

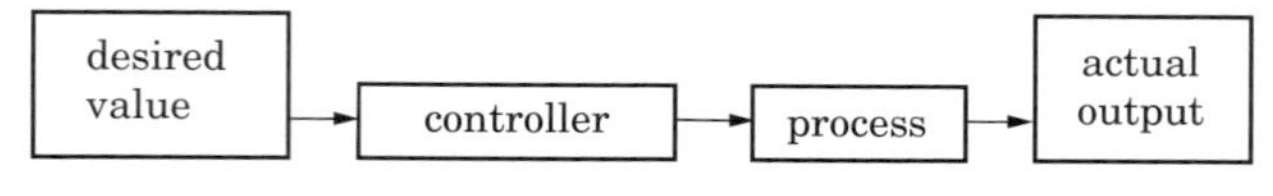

Fig. 21　Open-loop control system

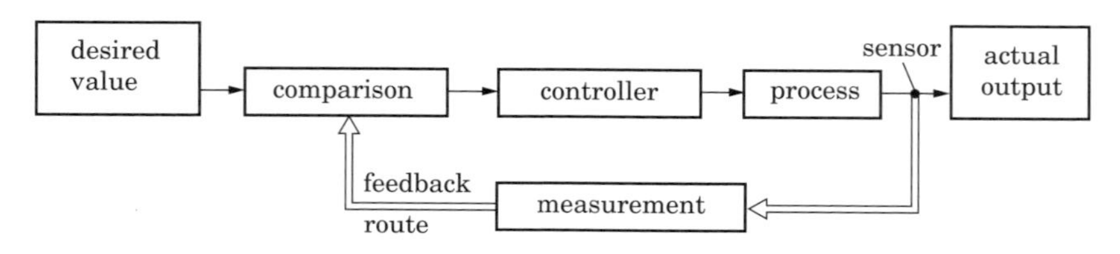

Fig. 22　Closed-loop control system

語 句

open-loop control system 開回路制御システム **utilize**[jú:tilaiz] 動利用する＝use **desired value** 目標値＝set poin **closed-loop control system** 閉回路制御システム＝feedback control system **deviation**[di:viéiʃən] 名偏差 **output value** 出力値 **manipulatable target** 操作対象 **input**[ínput] 名 動入力(する) **route**[ru:t] 名道すじ **sensor point** 測定点 **compare**[kəmpέə] 動比較する **feedback**[fí:dbæk] 名フィードバック **steer**[stiə] 動操縦する **driver**[dráivə] 名運転手 **sight**[sait] 名視覚，景色 **destination**[destinéiʃən] 名目的地 **steering wheel**(車などの)ハンドル **household**[háushould] 名家族，世帯 **appliance**[əpláiəns] 名器具 **refrigerator**[rifrídʒəreitə] 名冷蔵庫 **knob**[nɔb] 名柄，(ドアなどの)にぎり

●イントロ● フィードバックということばは日本語化している．その正確な意味について英文を読みながら考えてみよう．

構文

▶本文の1行目から4行目まで➡開回路制御システムと閉回路制御システムを対比させて述べた文章．(→解説)

▶by comparing actual output value with…➡ compare A with B は「A を B と比較する．」「出力値を入力した目標値と比較することによって得られた偏差を利用する」の意．

▶The route from the sensor point, where the output value is measured, to the controller…➡ この from と to とは関連している．…, where…measured, は関係副詞の限定用法だが文脈をわかりやすくするためにカンマ（，）で囲んである．「制御された出力が測定されるセンサ測定点から調節器までの道筋は…点と呼ばれる」．

▶This route, which can only transmit information, is…➡ 「情報だけを伝達するこの導線は….」情報だけと断っているのは動力を伝えるための導線ではないからである．

解説 機器，特に電気機器を最も望ましい状態で使用するために，以前は人間がハンドルを回して操作したが，現在では機械自身がコントロールするようになった．機械がコントロールする装置を自動制御という．

本文に述べたように，自動制御にはフィードバック回路のあるものとないものとがある．電子レンジや洗濯機は前もって加熱時間やすすぎの時間などをボタンを押して設定しておく．このように設定された順序どおりに作動するものをシーケンス制御（sequential control）といい，フィードバック回路をもっていない．これに対してエアコンでは，最初に目標値（desired value）を設定しておくと，それ以後は外界の温度をセンサが感知して，目標値に一致させるように自動的に動作する．このような制御をフィードバック制御（feedback control）といい，シーケンス制御より進んだ形式の制御といえる．この制御機器はもちろんフィードバックの回路をもっている．

問題

1. 次の（　）の中の語を並べ変えて正しい英文とし，それを日本語に訳せ．
   (1) A closed-loop control system (a sensor, has, a feedback route, and).
   (2) The route from the sensor point to the controller (called, a feedback route, is).
   (3) A feedback route (transmit, only, can, information).
   (4) A closed-loop control system (has, a refrigerator, applied, in, been).
2. 本文の内容と合っているものには○，合わないものには×をつけよ．
   (1) An open-loop control system is similar to a closed-loop control system.
   (2) A closed-loop control system can drive a car like a human being.
   (3) A thermostat is a sensor used to measure the actual temperature.
   (4) Few electrical appliances have a closed-loop control system.

5 NC工作機械

NC Machine Tool

NC (Numerically Controlled) machine tool is automatically controlled by means of discrete numerical values stored on a magnetic tape, or various type of discs, etc. In the case of NC machine tool, such conditions as relative location between a workpiece and tool, traveling path of a tool, feeding speed of a table, revolution speed of a main shaft are designed and stored on a memory medium, and the memorized information is transformed into a series of command signals. These signals are transmitted to a servomechanism, which directs a movable device to move as specified by the designer.

The NC machine tool equipped with a smart microprocessor and high density memory is called a computer-aided numerically controlled machine tool (CNC machine tool).

One of the most important advantages of this system is the tremendous versatility in product manufacture, as opposed to simply increasing the number of products. Because market requirements have recently shifted from "mass production of a few types" to "small production of various types," such CNC machine tool has become dominant in many manufacturing facilities.

語句

numerically[nju:mérikəli] 副数値によって　**automatically**[ɔ̀:təmǽtikəli] 副自動的に　**discrete**[diskrí:t] 形分離している　**magnetic tape** 磁気テープ　**traveling path** 移動径路　**feeding speed** 送り速度　**memory medium** 記憶媒体　**command signal** 命令信号　**servomechanism**[sə́:vouméкənizm] 名サーボ機構　**specify**[spésifai] 動明記する　**smart**[smɑ:t] 形気のきいた，利口な　**microprocessor**[máikrɔprəsesə] 名マイクロプロセッサ　**density**[dénsiti] 名密度　**computer-aided numerically controlled machine tool** コンピュータ支援数値制御工作機械　**tremendous**[triméndəs] 形ものすごい　**shift**[ʃift] 動移し変える　**mass production of a few types** 少品種大量生産　**small production of various types** 多品種少量生産　**dominant**[dɔ́minənt] 形支配的な，有力な

●イントロ●　NC 工作機械が一般的になってきたが，最近は一層進歩して CNC 工作機械になってきている．

▶**discrete numerical values stored on…**➡「…の上に記録されている離散型の数値によって」.（→解説）

▶**the memorized information is transformed into…**➡「記録されている情報は一連の命令信号に変換される．」この signal は電気信号のこと．

▶**a servomechanism, which directs a movable device to move…**➡「これら信号がまずサーボ機構に伝達され，そのサーボ機構が設計者が決めたように可動部分を動作させる」.（→解説）

▶**One of the most important advantages of this system**➡「このシステムの最も重要な利点の一つ」．最も重要な利点が複数になっているのは不合理な表現だが，英語でよく使われる表現である．

▶**as opposed to…**➡「製品の数を単純に増加させる（従来の）方法とは違って」.

▶**CNC machine tool has become dominant**➡「CNC 工作機械は…の中で優位を占めている．」現在完了の継続の用法．

解説

サーボ機構（Servomechanism）　サーボ機構は，ロボットの腕の運動や工作機械の刃物の位置の制御のように，力学的な位置の制御のための機構であり，当然のことながら運動を起こさせる元になる，強力でしかも正確な動力（actuator）の開発が必要になる．拡大機構を含むことが多く，運動の源泉をサーボモータ，ソレノイドなどに求めることが多い．これに対して，化学工場などで行われる濃度や温度などの制御は**プロセス制御（process control）**と呼ばれる．

問題

1. 次の（　）の中の語を並べ変えて正しい英文とし，それを日本語に訳せ．
   (1) NC in "NC machine tool" (for, the acronym, is, numerical control).
   (2) NC machine tool (not, but, has, the machine tool, only, also, a computer).
   (3) In an old NC machine tool, the numerical values (stored, in, were, a magnetic tape).
   (4) Nowadays, almost all NC machine tools (controlled, are, computer numerical).
2. 本文の内容と合っているものには○，合わないものには×をつけよ．
   (1) A NC machine tool is operated by programmed commands.
   (2) CNC is an acronym for Computer Numerical Controlled.
   (3) One of the most important advantages of the CNC system is its productivity.
   (4) Market requirements have shifted to "small production of various types."

コラム in case of と in the case of の違い

「～の場合は」として in case of と in the case of の二つには明確な使い分けがある．後者の in the case of が「(何か特定の誰か・何か）の場合は」という意味で，日本語で頻繁に使うニュアンスに近い．in case of では，「(何か異常・緊急事態）が起こった場合は」という，忠告や提案をする場合に用いられる．下記に例を挙げる．

in case of emergency「緊急事態の場合は」
in case of fire「火災が起きた場合は」
in case of an earthquake「地震に備えて」
just in case「念のために」

6 キャム（コンピュータ支援製造）

CAM

CAM is an acronym for "computer-aided manufacturing," and is closely related to CAD. In a traditional factory, jobs and tasks were arranged by a foreman in charge of the shop. For instance, when receiving the design drawing for work, he had to arrange machines to be used, materials to be fed, workers to be alloted, time schedule for procedures, and so on.

Nowadays, such time-consuming and troublesome work can be replaced by a computer loaded with sophisticated software. This sort of computerized system is called CAM. The newest CAM systems, in addition to the above-mentioned jobs, can perform personnel control. Supposing a worker suddenly takes a day-off, a foreman's one touch on a computer keyboard displays the most suitable rearrangement of personnel to cause the least disruption of the manufacturing schedule. The benefits brought about by CAM include increased productivity, significant reduction of wasteful time, and improved quality of products.

語　句

job[dʒɔb] 名仕事，地位　**task**[ta:sk] 名作業　**foreman**[fɔ́:mən] 名職長　**in charge of…** を監督している　**allot**[əlɔ́t] 動配分する，充当する　**procedure** [prəsí:dʒə] 名行動，手順　**nowadays**[náuədeíz] 副この頃では　**time-consuming** 時間のかかる　**troublesome**[trʌ́blsəm] 形厄介な　**in addition to…** に加えて　**above-mentioned** 前に述べた　**personnel**[pə:sənél] 名形職員（の）　**keyboard** [kí:bɔ́:d] 名（ピアノ，ワープロなどの）鍵板　**rearrangement**[rí:əréindʒmənt] 名再配置　**disruption**[disrʌ́pʃən] 名破壊　**reduction**[ridʌ́kʃən] 名減少← reduce 減少する　**wasteful**[wéistfəl] 形むだな

●イントロ●　CAD と CAM とは一体になって CAD/CAM になりつつある．工場や会社の経営は CAD と CAM とがなくては成り立たなくなってきている．現在，市場にある CAM ソフトの多くは CAD/CAM の構成になっている．

▶**a foreman in charge of the shop**➡「工場を監督している職長」.

▶**when receiving the design**…➡ receiving は分詞構文で，それに接続詞 when が付加された形.「仕事についての設計図面が渡されたとき，彼は～をしなければならなかった」.

▶**This sort of computerized system is called**…➡「このようなコンピュータ化されたシステムは…と呼ばれる」.

▶**in addition to the above-mentioned jobs** ➡「前述の仕事に加えて」.

▶**Supposing a worker suddenly takes**…➡ supposing は条件を表す分詞構文.「労働者が突然休暇を取ったと考えると」.

▶**The benefits brought about by CAM** ➡ bring about は「もたらす，引き起こす」の意.「CAM によってもたらされる利益は」.

問 題

1. 次の（　）の中の語を並べ変えて正しい英文とし，それを日本語に訳せ.

（1） A foreman (arranged, machines, materials, to be used, for, had, and, to be fed).

（2） CAM (can, which, work, in place of, is, a computerized system, a foreman).

（3） The benefits brought about by CAM (increased, productivity, include).

（4） A foreman's one touch (suitable, displays, rearrangement, the most, of personnel).

2. 本文の内容と合っているものには○，合わないものには×をつけよ.

（1） CAM is an acronym for "computer-aided manufacturing."

（2） CAM is the same as CAD.

（3） CAM is closely related to CAD.

（4） The newest CAM system can perform personnel control.

7 コンピュータ統合生産システム

CIM

CIM (computer-integrated manufacturing) system is a system in which all manufacturing-related functions and informations are connected in a computerized network in order to optimize manufacturfing activity. Specifically, functional areas such as design, analysis, planning, purchasing, cost accounting, inventory control, product distribution, and material management are organically linked with each other through computerized network to obtain the maximum efficiency.

Recently, CIM system also includes such new fields as market research, management strategy, prediction of economic efficiency, and decision making on management policy.

In the CIM factory of the future, no part nor finished products will even be standing or waiting ; that is each part or the finished either is being worked on or is on its way to the next work station or destination. In conventional factories, it is said that a part is processed only about 5% of the time. In future, there shall be no storage of works in process and no warehousing in the factory. To accomplish this, parts are preferably transferred by robots instead of human workers. CIM, excluding sales and management business, is sometimes called factory automation (FA) in Japan.

語　句

optimize[ɔ́ptimaiz] 動最適化する　**specifically**[spesífikəli] 副明確に，特別に　**functional**[fʌ́ŋkʃənəl] 形機能の，便利な　**purchasing**[pə́:tʃəsiŋ] 名購買　**cost** [kɔ́:st] 名代価，費用　**accounting**[əkáuntiŋ] 名会計　**inventory**[ínvəntri] 名商品目録　**organically**[ɔ:gǽnikəli] 副有機的に　**market research** 市場調査　**management strategy** 経営戦略　**prediction**[pridíkʃən] 名予報，予言　**decision** [disíʒən] 名決定←decide 決定する　**destination**[destinéiʃən] 名目的地　**accomplish**[əkɔ́mpliʃ] 動完成する　**exclude**[iksklú:d] 動除外する⟷include 包含する　**factory**[fǽktəri] 名工場　**automation**[ɔ:təméiʃən] 名オートメーション，自動化　**warehouse**[wɛ́əhaus] 名動倉庫（に入れる）

●イントロ●　CIMはさきに述べたCADやCAMのほかに，さらには検査，販売，営業，方針決定などを含めた会社全体の組織である．会社の統制部門と工場の現業部門とをコンピュータを介して結び，リアルタイムで情報を交換し，能率的に経営を行うものをいう．

CAD，CAM，CIMとまぎらわしいが，それがどのような英語の省略かを知っていれば混同することはない．

構文

▶**a system in which all manufacturing-related functions andinformations are connected**…➡which の先行詞は system．「CIMとは，すべての生産関連の機能と情報とが…とつながっているシステムである」．

▶**functional area such as…are organically linked each other**➡「…のような機能的分野がお互いに有機的に結合している」．

▶**no part nor finished products will even be standing or waiting**➡no と nor とは関連している．「一つの部品も一つの完成品も停止したり待機したりすることはないであろう」．

▶**…is on its way to～**➡「…は次の作業場か目的地へ送られる途中にある」．

▶**by robots instead of human workers instead of**…➡insted of は「…の代わりに」の意．「(部品）は人間労働者によらずにロボットによって輸送されることが望ましい」．

問　題

1. 次の（　）の中の語を並べ変えて正しい英文とし，それを日本語に訳せ.
   （1） CIM (to, optimize, a system, is, manufacturing activity).
   （2） All informations (connected, in, are, a computerized network).
   （3） Functional areas (linked, organically, are, each other, with).
   （4） A robot (role, in, an important, plays, the CIM factory).
2. 本文の内容と合っているものには○，合わないものには×をつけよ.
   （1） CIM is an acronym for “computer-integrated manufacturing.”
   （2） CIM, CAD, and CAM are the same meanings.
   （3） FA is almost the same as CIM in Japan.
   （4） In future, there will be no warehousing in the factory.

コラム　ハイフンの使いどころ

ハイフン（hyphen）またはダッシュ（dash）には種類があり，それらの用途は一つずつ異なる．ハイフンの役割の多くは，二者を繋げるものであるが，その繋げられる対象によって，使い分けられる.

ハイフン（-）：一つの単語を（改行の為に）分けたり，二つ以上の単語を繋いで複合語を作ったり，二つ以上の単語で後の名詞を修飾する際に用いる．パソコンのキーボードなどでも簡単に打てるものである.

エンダッシュ（en dash, –）：範囲や期間を表すときや，ハイフンで繋げられた複合語を繋ぐときに用いる．和文で範囲を表すとき「～（tilde)」が使われるが，英文では一般的ではなく，数式中で漸近的に等しい場合に用いられる．ハイフン二つ分ぐらいの長さで，特殊文字の部類に入る場合が多い.

10–20 m ➡ 10 m から 20 m までの範囲を示す．スペースは不要.

10 − 20 ➡「10 引く 20」を表す場合は，両端にスペースを付けたマイナス記号（en dash 記号と似ている）を用いる.

$\left(1+\frac{1}{n}\right)^n \sim e$ ➡英文でのチルダ（~）は数式と同様の扱い.

エムダッシュ（em dash, —）：文と文を繋げたり，追加情報などのフレーズを協調的に挿入したりするときに用いる．エンダッシュと同様にスペースは不要．エンダッシュより長い記号.

8 フレキシブル生産システム

FMS

FMS (flexible manufacturing system) is a sort of a system by which various kinds of products can be alternatively manufactured on the same manufacturing line by the aid of a computer. The flexibility in manufacturing is now considered to be as important as the efficiency, and has become a crucial factor to survival under fierce competition for market share, which demands that a factory always responds quickly to market needs. The traditional factory equipped with a rigid assembly line, and rigid organization eventually cannot satisfy society's requirements. If FMS is ideally adopted, each product on the line will be of a different type; that is, any two products are not identical.

FMS used in mechanical manufacturing usually involves CNC machine tool, automatic assembling machines, automatic inspectors, and robots. The line equipped with these computerized devices can change the type of products just by adopting another software. In 1970, when FMS was first introduced, this system was used mainly in machine shops, but it has now spread to various fields such as the clothing, foods, and petroleum industries.

語　句

flexible[fléksəbl] 形柔軟な，素直な　**alternative**[ɔ:ltɔ́:nətiv] 形二者択一の　**efficiency**[ifíʃənsi] 名効率，能率　**crucial**[krú:ʃiəl] 形決定的な，重大な　**survival**[səváivəl] 名生き残ること（人）　**fierce**[fiəs] 形激しい　**competition**[kɔmpitíʃən] 名競争，試合← compete 競争する　**share**[ʃεə] 動分配する 名割当て　**demand**[dimá:nd] 動要求する　**respond**[rispɔ́nd] 動応答する　**eventually**[ivéntʃjuəli] 副結局　**identical**[aidéntikl] 形同一の　**spread**[spred] 動広げる　**clothing**[klóuðiŋ] 名衣服

●イントロ●　将来ますます消費者のニーズは多様化する．それにこたえるための生産ラインがこの FMS システムである．

構文 ▶**has become a crucial factor to survival under fierce completition** ➡ has become は現在完了継続.「フレキシビリティは厳しい競争下にあって生き残るための絶対的な要件になっている」.

▶**…, which demands that～** ➡ which の先行詞はやや離れている flexibility.「フレキシビリティは工場が市場の要求にすばやく対応できることを必要とする」.

▶**If FMS is ideally adopted, each product will…** ➡ If…は仮定法の用法.「もしも FMS が理想的に採用されるならば，各製品は…であろう」. will は仮定法に対応して用いられた未来の助動詞.

▶**…be of a different type** ➡「ライン上の各製品は異った形態のものになる」.

解説 人は誰でも他人より優れているものを望むと同時に，同一のものをもちたがらない傾向がある．そのような要望が市場に反映すれば，工場で作られる製品ができるだけ多種多様であることが望ましい．これは，今まで最も望まれていた低価格中心の大量生産と相反する方式である．FMS が現在どこの企業でも話題になっているのはこの理由による．これから人間の価値観が多様化するにつれて，FMS がますます求められていくのは確実である．

問 題

1. 次の（ ）の中の語を並べ変えて正しい英文とし，それを日本語に訳せ.
   (1) The traditional factory (with, a rigid assembly line, equips).
   (2) FMS is a manufacturing system in which (is, some, amount of, there, flexibility).
   (3) The line (these, with, computerized devices, equips).
   (4) This system (introduced, in, was, first, machine shops).
2. 本文の内容と合っているものには○，合わないものには×をつけよ.
   (1) FMS produces various kinds of products on the same manufacturing line.
   (2) CNC machine tool is necessary for the FMS used in mechanical manufacturing.
   (3) It is difficult to adopt the FMS in machine shops.
   (4) The traditional factory has the advantages of responding quickly to market needs.

コラム 便利な動詞 determine

「～を決定する」「(事実・原因など) を究明する」「～を測定する」などの意味を多彩にもつ determine は技術英文でもよく使われる便利な動詞である.

A meetings is held to determine standards. (基準決定会議が開催される.)

This study determines the flow structure. (本研究で流動構造を調査する.)

The number of particles was determined by SEM. (SEM (scanning electron microscope：走査型電子顕微鏡) で粒子数を測った.)

9 燃料電池

Fuel cell

A fuel cell is an electrical generator. Unlike ordinary electrical generators, the fuel cell generates electricity by a chemical reaction. The fuel cell generates electricity on the principle that is a reverse of "electrolysis of water". Water is decomposed into hydrogen and oxygen by electric current. This is reversed in the fuel cell, and the reaction between hydrogen and oxygen produces water and electricity. The fuel cell consists of two electrodes called the anode and the cathode, an electrolyte that carries electrically charged particles from one electrode to the other, and a catalyst. A single fuel cell generates a small amount of electricity. Therefore, many fuel cells are usually assembled into a stack for practical use.

There are several types of fuel cell. They are classified by the electrolyte they employ into alkali fuel cells (AFC), molten carbonate fuel cells (MCFC), phosphoric acid fuel cells (PAFC), solid oxide fuel cells (SOFC), polymer electrolyte fuel cells (PEFC). The PEFC is currently used for home fuel cell systems and fuel cell-powered cars.

The main advantage of the fuel cell is that it generates electricity with very little pollution; there is no polluting exhaust gas, only a harmless by-product water. A hydrogen station has been prepared for a fuel cell-powered car. If the efficiency of the fuel cell improves, the fuel cell will come to be used more widely.

語句

fuel cell[fjúːəl sél] 名燃料電池　**reverse**[rɪvˊəːs] 形逆の　**electrolysis of water** 水の電気分解　**electrode**[ɪléktroʊd] 名電極　**alkali fuel cells (AFC)** アルカリ型燃料電池　**molten carbonate fuel cells (MCFC)** 溶融炭酸塩型燃料電池　**phosphoric acid fuel cells (PAFC)** りん酸型燃料電池　**solid oxide fuel cells (SOFC)** 固体電解質型燃料電池　**polymer electriolyte fuel cells (PEFC)** 固体高分子型燃料電池　**home fuel cell systems** 家庭用燃料電池　**by-product** 副産物

●イントロ●　燃料電池は水の電気分解の逆の反応を利用した新しい発電方法であり，火力発電などと異なり汚染ガスを出さないので，クリーンなエネルギーとして，将来の発展が望まれる分野の一つである．特に最近は，家庭用燃料電池を使用したコージェネレーションや燃料電池自動車として実用化されている．また，燃料電池自動車の普及に向けて水素ステーションなどがガソリンスタンドに設置されている．

構文 ▶**The fuel cell generates electricity on the principle that is a reverse of "electrolysis of the water".** ➡「燃料電池は『水の電気分解』の反対の原理によって電気を生み出す」．

▶**many fuel cells are usually assembled into a stack** ➡「たくさんの燃料電池が普通スタック（積み重ねられたもの）へと組み立てられる」．

▶**The main advantage of the fuel cell is that ….** ➡ that の関係代名詞ではなく that 節の that.「燃料電池の主な利点は…ということです」．

▶**only a harmless by-product water** ➡「無害の副産物である水だけ」．

問題

1. 次の（　）の中の語を並べ変えて正しい英文とし，それを日本語に訳せ．
   （1） The fuel cell (generates, a chemical reaction, by, electricity).
   （2） Electricity (is, between, produced, hydrogen, by, the reaction, and oxygen).
   （3） Many types of fuel cells (have, developed, been).
   （4） AFC (was, named, for, the famous spacecraft, *Apollo*, used).
2. 本文の内容と合っているものには○，合わないものには×をつけよ．
   （1） A fuel cell turns a turbine to generate electricity.
   （2） A fuel cell burns hydrogen to generate warmth.
   （3） The home fuel cell system has begun to spread.
   （4） A fuel cell-powered car does not exhaust any gas.

10 太陽電池

Solar cell

A solar cell is sometimes called a solar battery; however, it is not a battery but an electrical generator. The solar cell is different from the common battery in that electricity is not saved. The solar cell is a device that converts light energy into electricity using the photovoltaic effect. The solar cell is composed of semiconductors such as silicon. Photovoltaic power generation is the generation method that produces electricity as long as light of the sun can be obtained.

The solar cell does not need fuel like thermal power generation and nuclear power generation, and it does not produce any exhaust gas. In this sense, the solar cell is regarded as a kind of a clean energy that is friendly to the global environment. In addition, the solar cell is very reliable because there are no operation parts such as an engine and turbine. In short, the solar cell has many advantages over other power generation technologies.

One aspect to note is that the electricity generated by a solar cell is not always steady. The electricity generated by the photovoltaic effect is dependent on the amount of light. No electricity is generated by the solar cell when there is no light. Therefore, the solar cell is sometimes accompanied by a battery to store electricity generated by the solar cell in the daytime and to use the electricity at night.

語　句

solar cell[sóulər sél] 名太陽電池　**photovoltaic effect** 光電効果　**power generation** 発電機　**reliable** [riláiəbl] 形信頼できる　**turbine**[tə́:rbin] 名タービン　**aspect** [ǽspekt] 名面，見地，外観　**steady**[stédi] 形一定である，安定した　**daytime**[déitàim] 名 昼間

●イントロ●　太陽電池は半導体による光電効果を利用した発電装置であり，タービンを回して発電する火力発電などと発電原理が根本的に異なる新しい発電方

法である．また，太陽電池は太陽光から電気を生み出すので，火力発電や電子力発電などとは異なり燃料を必要とせず，また汚染された排気ガスを出さないので，クリーンなエネルギーである．初期には電卓や腕時計などに使用されていたが，最近は家屋の屋根やビルの屋上などに設置され，太陽電池により発電された電気も使用されつつある．

構文 ▶**The solar cell is different from the common battery in that electricity is not saved.** ➡「太陽電池は電気が保存できない点で普通の電池とは異なる」．

▶**The solar cell is a device that converts light energy into electricity using the photovoltaic effect.** ➡ which の先行詞はその前の a device「太陽電池は光電効果を使って光のエネルギーを電気に変換する装置である」．

▶**In this sense, the solar cell is regarded….** ➡「この点で，太陽電池は…と見なされる」．

▶**One aspect to note is that….** ➡「注意しなればならない一つのことは…ということです」．

問　題

1. 次の（　）の中の語を並べ変えて正しい英文とし，それを日本語に訳せ．
   (1)　A solar cell (by,　generates,　electricity,　the photovoltaic effect).
   (2)　A solar cell (is,　semiconductors,　made,　of,　silicon,　such as).
   (3)　A solar cell (are,　is,　no moving parts,　very reliable,　because,　there).
   (4)　A solar cell (cannot,　be,　sufficient,　without,　light,　used).
2. 本文の内容と合っているものには○，合わないものには×をつけよ．
   (1)　A solar cell is a battery to store electricity.
   (2)　A solar cell produces so-called clean energy.
   (3)　A solar cell can be used for various applications.
   (4)　A solar cell generates electricity by heating the solar cell.

11 ハイブリッド車

Hybrid car

A hybrid car is a vehicle that uses two different types of power source, such as an internal combustion engine and an electric motor. The hybrid car is different from other cars in that the hybrid car has a motor. There are three types of hybrid car; the series type, the parallel type, and the series-parallel type. In the series type, the engine is only used for turning a generator to generate electricity, and the motor is used as the power source to move the tires. On the other hand, both the engine and motor are used as the power source to move the tires in the parallel type. The motor is used as the power source to move the tires while the battery supplies electricity, but the motor turns into the generator and the engine is only used as a power source to move the tires when the battery cannot supply electricity. In the series-parallel type, the motor is always used as the power source to move the tires because the car is equipped with a generator. The car runs by either only a motor or both a motor and an engine according on the actual driving environment.

One merit of the hybrid car is its fuel-efficiency. The hybrid car makes use of efficiency-improving technologies such as regenerative brakes. Since the hybrid car is fuel-efficient, it emits little exhaust such as carbon dioxide or nitrogen oxide. Therefore, the hybrid car is less harmful to the environment and is called an eco-car. At present, the hybrid car is higher in price than the gasoline car. When the price of the hybrid car becomes comparable to the gasoline car, hybrid cars will be driven more.

語句

hybrid car[háibrid ká:r] 名ハイブリッド車　**vehicle**[ví:ikl] 名乗り物，車　**power source** 動力源　**supply**[səplái] 動供給する　**regenerative brake** 回生ブレーキ　**fuel-efficient** 省燃費

●イントロ●　ハイブリッド車は燃費がよく，排気ガスが少ない環境にやさしい自動車である．ハイブリッド車はモータとエンジンの両方を積んでいるが，その駆動形式はさまざまである．ハイブリッド車は電気自動車，燃料電池自動車などと一緒にエコカーと呼ばれ，環境を守るためにその普及が進められている．

構文　▶**A hybrid car is a vehicle that uses two different types of power source,** ➡ that の先行詞はその前の a vehicle「ハイブリッド車は二つの異なった種類の動力源を使う乗り物です」．

▶**In the series type, the engine is only used for turning a generator to generate electricity.** ➡「シリーズ型では，エンジンは電気を作るため発電機を回すためだけに使われる」．

▶**The hybrid car makes use of efficiency-improving technologies such as regenerative brakes.** ➡「ハイブリッド車は回生ブレーキのような効率を向上する技術を使っている」．

問題

1. 次の（　）の中の語を並べ変えて正しい英文とし，それを日本語に訳せ．
   (1) The hybrid car (both, has, and, a combustion engine, an electric motor).
   (2) The hybrid car (drive, can, more miles, other cars, than).
   (3) Regenerated brakes (to, generate, are, brakes, electricity).
   (4) The hybrid car (is, expensive, without a motor, than, more, comparable cars).
2. 本文の内容と合っているものには○，合わないものには×をつけよ．
   (1) In all hybrid cars, the engine is used as the power source to movie tires.
   (2) The motor plays an important role in all hybrid cars.
   (3) The hybrid car is classified as an eco-car.
   (4) The hybrid car is becoming popular due to its high fuel efficiency.

12 超伝導

Superconductivity

Superconductivity is a phenomenon occurring at very low temperatures in many metals or metal oxides, in which DC electric resistance becomes zero at a transition temperature inherent to the material. In 1911, H. K. Onnes discovered superconductivity while studying the variation in electric resistance of mercury in relation to temperature. He observed that the resistance dropped sharply to a very low, nearly immeasurable value at a temperature of 4.2K. In 1957, J. Bardeen, L. N. Cooper, and J. R. Shrieffer constructed the first successful theory of superconductivity, by which the rearrangement of electrons allowing superconductivity was explained.

In 1986, it was found that very interesting metal oxide ceramics had a transition temperature as high as 35 K. Thereafter successive studies have found substances with a transition temperature higher than 130 K. These new metal oxide superconductive materials can be very useful, because we can cool them with cheaper coolants, such as liquid nitrogen instead of expensive liquid helium. Making use of the immense magnetic power that can be generated in a zero-resistance current in superconductive materials, a linear-motor car, a super collider, a superconducting quantum interference device (SQUID) magnetometer, and magnetic resonance imaging (MRI) have been developed.

語 句

superconductivity[sjù:pəkɔndʌktíviti] 名超伝導　**DC** 直流＝direct current（AC 交流）　**transition**[trænsíʒən] 名変化，移り変り　**transition temperature** 転移温度　**inherent**[inhíərənt] 形本来備わっている　**variation**[vɛəriéiʃən] 名変化← vary 変化する　**immeasurable**[iméʒərəbl] 形無限の　**explain**[ikspléin] 動説明する　**ceramics**[sirǽmiks] 名陶磁器類（単数扱い）　**linear-motor car** リニアモータカー　**collider**[kəláidə] 名衝突型加速器（超高速粒子を衝突させて物質の微細構造を研究する機器）

●イントロ● 超伝導の話題はこのところ落ち着いた感じだが，その可能性に対する期待は依然として高く，将来の発展が望まれる分野の一つである．この超伝導材料は高性能な磁力計（SQUID）やリニアモータとして実用化されている．

▶**…, in which DC electric resistance becomes zero…**➡ which の先行詞はその前の metals and metal oxides.「それら金属および金属酸化物の中では，直流抵抗がその物質に特有の転位温度でゼロになる」.

▶**in relation to temperature** ➡「温度に関係して」.

▶**by which the rearrangement of electrons allowing superconductivity was explained.** ➡ which の先行詞はその前にある theory. 転移点の前後で原子の配列が変わり，それにつれて当然電子の配置が変わってくる.「その理論によって超伝導を可能にするような電子の再配置が説明された」.

▶**making use of the immense magnetic power that…**➡「…によってもたらされる強大な磁力を利用して」.

問 題

1. 次の（ ）の中の語を並べ変えて正しい英文とし，それを日本語に訳せ.
   (1) The electric resistance of the superconductor (becomes, at, zero, low temperatures).
   (2) This phenomenon (has, more than, been, known, for, a hundred years).
   (3) A linear motor car (able, will, run, 40 minutes, Tokyo and Nagoya, be, to, in, between).
   (4) MRI is a superconducting device (to, of the human body, diagnose, conditions).
2. 本文の内容と合っているものには○，合わないものには×をつけよ.
   (1) A superconductor is a metal or a metal oxide.
   (2) A superconductor can be used at zero degrees Celsius.
   (3) A superconductor needs to be cooled by ice or dry ice.
   (4) A linear motor car uses a superconducting magnet.

○● 章末問題 ●○

1.　空欄を埋めて，日本語に訳せ.

(1)　CAD is an acronym (　　　) computer-aided design.

(2)　The word, mechatronics, was first used (　　　) Japan.

(3)　A sensor is a device (　　　) detect something.

(4)　A feedback control system uses a controller (　　　) obtain the desired value.

(5)　CNC is a NC machine tool (　　　) a computer.

(6)　CAM is closely related (　　　) CAD.

(7)　CIM is a system in which all informations are connected (　　　) a network.

(8)　FMS can manufacture various products (　　　) the aid of computer.

(9)　A fuel cell generates electricity (　　　) a chemical reaction.

(10)　A solar cell does not need fuel (　　　) thermal power generation.

(11)　A hybrid car has not only an engine (　　　) also a motor.

(12)　Superconductivity occurs (　　　) a very low temperature.

2.　次の文を英語に直せ.

(1)　CAD とは製品の設計のためにコンピュータを使うことをいう.

(2)　メカトロニクスはいわゆる和製英語である.

(3)　リミットスイッチは一種のセンサです.

(4)　フィードバック制御システムは冷蔵庫の温度調整に使われている.

(5)　CNC 工作機械が多くの製造工場で優位を占めている.

(6)　CAM はコンピュータ支援製造の頭字語です.

(7)　CIM は CAD や CAM とは少し異なります.

(8)　FMS は同じ製造ラインで異なった種類の製品が製造できるシステムです.

(9)　燃料電池は水素と酸素の化学反応で電気を発生させます.

(10)　太陽電池は半導体の光電効果を利用して発電する装置です.

(11)　ハイブリッド車は燃費がよいのでエコカーと呼ばれます.

(12)　超伝導磁石がリニアモータに使用されています.

V

管理技術

Managerial Engineering

かつて，会社の経営や工場の運営は，多くその立場にある者の長年にわたる経験や勘に頼っていた．その勘が冴えれば会社は発展し，逆に鈍れば会社が傾くこともあった．

しかし，現在は小企業は別として，少なくとも中規模以上の企業ではそのような経営は許されない．企業の管理あるいは経営に関する研究が進み，その成果を縦横に駆使した管理・経営が行われるのが常である．

それら管理・経営に関するいくつかのトピックスを読んでみよう．

1 互換性

Interchangeability

An automobile consists of thousands of parts, and the parts are often made in factories located hundreds of kilometers apart ; e. g. bodies are made in one factory, engines in another, and tires in yet another factory. It is, therefore, extremely important to make each part exactly the same in size, materials and performances. If they are properly made, when the hundreds of parts are brought together, they will be accurately assembled according to the blue print prepared in another place beforehand. In other words, these parts are manufactured with full interchangeability.

When the concept of interchangeability was weak, a worker had to file individual parts which did not fit to adjust for proper fitting. Now, the conditions in a well-managed factories have been entirely changed. All parts that pass an inspection department can always fit in the right place without any wasteful filing. It would not be overstatement to say that the concept of interchangeability has established the foundation of mass production which is supporting our prosperity at present.

語　句

interchangeability[íntəːtʃeindʒəbíliti] 名互換性　**locate**[lóukeit] 動（ある場所に）置く　**e. g.**[iː-dʒiː] たとえば＝for example　**bring together** まとめる　**beforehand** [bifɔ́ːhand] 副前もって　**weak**[wiːk] 形弱い　**file**[fail] 名動やすり（をかける）　**individual**[indivídjuəl] 形単一の，個人の　**fit**[fit] 動適合する（させる）　**wasteful** [wéistful] 形むだな　**overstatement**[óuvəstéitmənt] 名誇張　**support**[səpɔ́ːt] 動支持する　**prosperity**[prɔspériti] 名繁栄

●イントロ●　ゲージや誤差と密接にかかわる互換性という概念について述べている．工業製品では互換性を確保することで部品の利用性が高まる．この互換性部品生産により，フォード生産方式として知られる大量生産が可能となった．

しかし，現在ではこの互換性は部品に限らずソフトウェアなど幅広い分野で，

ある製品やデータを別のものに置き換えても問題なく使用できるという意味で使われることが多い．

▶**It is, there fore, extremely important to make…**➡ It は to 以下を表す形式主語．「各部品を…に作ることは極めて重要である」．

▶**If…, they will be accurately assembled** ➡ If…は仮定法の文で，they will…はそれを受ける結びの文．「もし…ならば，それら（部品）は…に従って正確に組み立てられるであろう」．

▶**It would not be overstatement to say that…**➡ would は if が使われていないけれども，仮定法的内容を受けている．It は to 以下を表す．「もしも…と言ったとしても言い過ぎではないであろう」．

問題

1. 次の（　）の中の語を並べ変えて正しい英文とし，それを日本語に訳せ．
   (1) An automobile (of, consists, thousands, of, parts).
   (2) It is extremely important (to, make, interchangeable, each part, exactly).
   (3) They (will, according to, assembled, be, accurately, the blue print).
   (4) A worker (must, to assemble, file, individual parts, them, properly).
2. 本文の内容と合っているものには○，合わないものには×をつけよ．
   (1) The concept of interchangeability used to be weak in previous times.
   (2) Interchangeability is not so important when there are many parts.
   (3) Interchangeability is sometimes troublesome.
   (4) The concept of interchangeability has led to the establishment of the mass production.

2 サンプリング

Sampling

It is very common in manufacturing that a lot of products are required to be inspected. If they are easily inspected, complete information about their properties may be gathered by 100% inspection. However, in cases where a vast number of products must be inspected or the inspection gives each product a little damage, we cannot carry out 100 % inspection. In these cases, sampling inspection is the most powerful and effective method.

Sampling inspection means that only a certain number of products are previously selected from a lot to be inspected and from the inspection we can statistically obtain satisfactory information about the whole population of the lot. If the sample gives a good result, we can naturally judge the target lot is of good condition, and thus the lot can be accepted.

When carrying out sampling inspection, the most important factor is the method of selecting samples from the lot. The method of selection is called random sampling, which means to choose the samples at random from the population without sampler's premeditated intention. For this random sampling, we use an icosahedron die or a table of random numbers.

語　句

gather[gǽðə] 動集める→ gathering 集合　**vast**[vɑ:st] 形非常に大きい　**sample**[sǽmpl] 動見本を集める　**sampling inspection** 抜取り検査　**powerful**[páuəful] 形強力な　**effective**[iféktiv] 形効果のある　**previously**[prí:viəsli] 副前もって　**select**[səlékt] 動選択する　**statistically**[stətístikəli] 副統計学に基づいて　**satisfactory**[sætisfǽktɔri] 形満足な　**population**[pɔpjuléiʃən] 名人口，母集団　**lot**[lɔt] 名ロット，商品などの一組　**random sampling** ランダムサンプリング，無作為抽出　**premeditated**[priméditeitid] 形計画的な　**icosahedron**[aikəsəhédrn] 名正二十面体　**icosahedron die** 乱数さい　**random number** 乱数

●イントロ●　日本の産業界が最も得意とする管理技術の一つに品質管理があ

る．その際にいかに正しくサンプリングするかが大きな問題になる．

構文 ▶**It is very common in manufacturing that…** ➡ It は that 以下を表す形式主語．「製造業において…ということは極めてあたりまえのことである」．

▶**100% inspection** ➡「全数検査」．これは sampling inspection「抜取り検査」に対することばで，製品全部を検査すること．

▶**in cases where** ➡「…の場合には」．where の後には節が来る．句が来る場合には in the case of（=in case of）を用いる．

▶**Sampling inspection means that…** ➡ that は接続詞．「抜取り検査とは，製品のうちの少数を前もって選び出し…することである」．

▶**When carrying out sampling inspection, …** ➡分詞構文で，「抜取り検査を実施するときには」．

問題

1. 次の（　）の中の語を並べ変えて正しい英文とし，それを日本語に訳せ．
 （1） Sampling inspection (is, powerful, the most, and, effective, method).
 （2） The inspection (a little, gives, each product, damage).
 （3） The most important factor (is, selecting, the method, samples, of, from the lot).
 （4） The method of selection (is, random, called, sampling).

2. 本文の内容と合っているものには○，合わないものには×をつけよ．
 （1） We have to choose good samples for sampling inspection.
 （2） Sampling inspection is a way to avoid conducting 100% inspection.
 （3） Random sampling is conducted by using a table of random numbers.
 （4） It is always the best way to inspect all the products.

3 検査と試験

Inspection and Testing

Inspection involves the formal checking of the properties or dimensions of materials and products while they are being made or immediately after they are completed. A complete inspection and testing program usually cover evaluation of the program design, conformance to design during manufacturing, and durability during subsequent storage, transport, and use. Parts rejected by inspection should be discarded. Measuring instruments also have to be inspected regularly, because imprecise instruments might result in the inaccurate inspection leading to shipping inferior products.

The terms inspection and testing are sometimes interchangeably used, but strictly speaking, inspection has a wider concept than testing. Inspection usually includes interpretation of specifications, measurements of products, methods of sampling, statistical analysis of results, standardization, and so on.

Inspection is usually handled by the inspection department of a shop. A room in which only inspectors work is an inspection room. This room is usually air-conditioned and sometimes also dust-controlled depending on the type of testing machines and objects to be tested.

 **語　句**

immediately[imí:diitli] 副ただちに　**evaluation**[ivæljuéiʃən] 名評価　**conformance** [kənfɔ́:məns] 名適合　**durability**[djuərəbíləti] 名耐久性　**storage**[stɔ́:ridʒ] 名貯蔵，保管　**transport**[trænspɔ́:t] 名動輸送（する）　**imprecise**[imprəsáis] 形不正確な　**inferior**[infíəriə] 形粗悪な，劣った

●イントロ●　検査あるいは試験は，かつて考えられていた概念よりもずっと広くなっており，会社や工場にとって極めて重要な部門である．

▶**while they are being made or** … ➡ while は when よりも長い時間経過を示し，その後に進行形が来ることが多い．「それらが製造されている間か，またはそれらが完成したすぐ後で」．

▶ **…, because imprecise instruments might result in the inaccurateinspection** ➡「というのは，不正確な（検査）器具は不良品の出荷を招くようないい加減な検査になってしまうおそれがあるからである」．

▶**strictly speaking** ➡「正確にいうと」．広い意味の分詞構文．

▶**standardization** ➡「標準化」．製品の規格を決めることだけではなく，製造工程，設計，検査，販売などについての基準を決めることをいう．

▶**This room is usually air-conditioned and sometimes also dust-controlled** ➡「この部屋は普通空調されていて，時々塵埃コントロールもされている」．空気中に浮遊するごみを極力少なくすることで，特にIC基板の製造で重視されるようになった．

問 題

1. 次の（　）の中の語を並べ変えて正しい英文とし，それを日本語に訳せ．
   （1） Inspection (is, the properties or dimensions, to, check, of products).
   （2） Measuring instruments (have, inspected, to, be, regularly).
   （3） Inspection (is, by, handled, usually, the inspection department).
   （4） The concept of inspection (is, the concept of testing, than, a wider concept).
2. 本文の内容と合っているものには○，合わないものには×をつけよ．
   （1） Inspection and sampling have the same purpose.
   （2） Inspection and testing have the same purpose.
   （3） Parts rejected by inspection should be inspected again.
   （4） Parts rejected by inspection should be returned to the production department.

4 管理図 Control Chart

This chart can be used to determine whether the manufacturing process is under a stable condition or if urgent action is needed to recover stability, and is one of the most crucial measures to complete the best quality control.

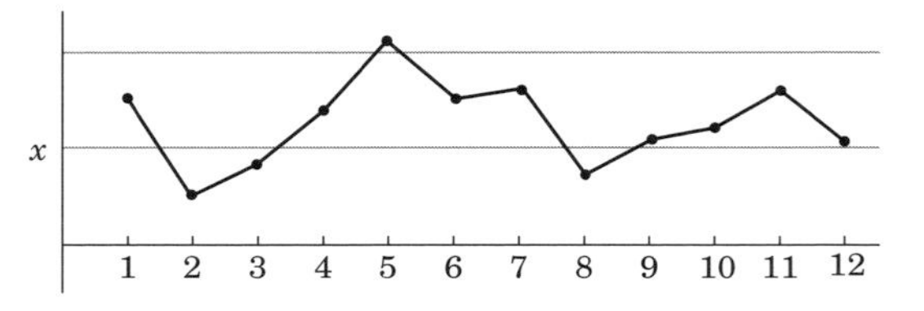

Fig. 23 $\bar{x}$ control chart (Example)

Usually, the chart consists of a central line and a pair of parallel lines called limit control lines because they represent the upper and lower limit of product characteristic. Measured values of characteristic such as dimension, temperature, and concentration are sequentially plotted on the chart. If these plots show a serious skewed trend or if they fall out of the two limit lines, the manufacturing process is judged to be operating under erroneous conditions such that inferior products might be finally produced.

The theory of control charts is based on statistics, so useful control charts must be designed on the basis of a very stable population. A population means a lot of products from which some samples are picked up for making the chart. The control chart is also called Shewhart chart because the theoretical background of the chart was proposed by Shewhart. This system was introduced by Dr. Deming to Japan after World War II, and has remarkably developed by the investigation in Japan.

語　句

chart[tʃɑ:t] 名図表　**stable**[steibl] 形安定した　**urgent**[ə́:dʒent] 形緊急の　**action**[ǽkʃən] 名活動　**recover**[rikʌ́və] 動取り戻す　**quality control** 品質管理＝QC　**parallel**[pǽrəlel] 形平行な　**upper**[ʌ́pə] 形上の方の⟷ **lower** 下の方の　**concentration**[kɔ̀nsentréiʃən] 名濃度，集中　**sequentially**[sí:kwənʃəli] 副連続して　**plot**[plɔt] 動点をつないで曲線を書く　**skewed**[skju:d] 形ゆがんだ　**trend**[trend] 名傾向　**erroneous**[iróuniəs] 形誤った　**theory**[θíəri] 名理論，学説　**background**[bǽkgraund] 名背景，経歴

●イントロ●　管理図法は品質管理の手法の中で最も有力なものの一つである．連続的に生産される品物を適当にサンプリングし，その特性値を図のようなチャートに打点し，打点の傾向から工程が適切か否かを判断する．

構文

▶**determine whether the manufacturing process is under a stable condition or if urgent action is needed** ➡ whether と if とはともにdetermineにつながる．ifは口語的に，whetherと同意に用いられる．「製造工程が安定した状態の下にあるか，または緊急な手段を講じる必要があるかどうかを判断する」．

▶**parallel lines called limit control lines because they represent** … ➡「それらは…を表わしているがゆえに限界線と呼ばれる 2 本の平行線」．

▶**under erroneous conditions such that inferior products might** … ➡「粗悪な製品が製造されるかもしれないような悪条件の下で」．

解説

最近の品質管理は製造工程よりももっと源流である設計段階における管理が問題になっている．製品の設計段階において，材料，工程，工作機械，工具，作業条件などを管理し，不良品が絶対に発生しないような管理方法が研究されている．品質管理技術の源流はアメリカの Shewhart であったのだが，それが日本で花開き，日本の工業製品の優秀さを生み出している．

問　題

1. 次の（　）の中の語を並べ変えて正しい英文とし，それを日本語に訳せ．
   (1) A control chart (used,　the manufacturing process,　is,　to,　evaluate).
   (2) A control chart usually (consists,　a pair of,　of,　a central line,　and,　parallel lines).
   (3) The dimensions of the manufactured products (plotted,　on,　are,　the chart).
   (4) The theory of control charts (is,　on,　based,　statistics).
2. 本文の内容と合っているものには○，合わないものには×をつけよ．
   (1) The control chart was developed in Japan.
   (2) The control chart was introduced and studied during World War II.
   (3) The control chart can be applied even if there are insufficient samples.
   (4) The control chart is one of the most important tools for quality control.

コラム　スラッシュの使いどころ

英文中のスラッシュ（slash，/）は or の意味で用いられ，選択肢や関連性の羅列を示す．「Dear Sir/Madam」「an on/off control circuit」．ただし，分数表現や年月日での使用は例外．和文では中点（bullet，・）を使うこともあるが，英文での中点は and を意味することに注意．また，スラッシュは語句を省略する場合にも使われ，without の略称で w/o（with は w/）はよく使用される．

5 オペレーションズリサーチ

Operations Research（OR）

Though it may sound strange, OR originated from the operational activity of army in England during the World War II. England was the target of bombing by German airforce, and it was so severe that England had to struggle to find ways to prevent air invasions. This anti-air raid study was the first step of OR. Now OR is defined as the application of scientific methods and techniques to determine the best solution when confronted with a difficult situation. In other words, it is a decision-making process carried out by means of mathematical aids where there are two or more alternative courses of action. This situation is figuratively expressed as "Seven unknowns and four equations."

The procedures needed for OR techniques are typically as follows：(1) formulate the target problems, (2) construct a model of the system, (3) select the most suitable technique, (4) obtain the best solution to the problem, and (5) implement the solution. There are numerous areas where OR has been applied. A familiar instance is encountered when deciding the optimal number and arrangement of elevators, staircases, or toilets in a big building or school. OR can provide the key technique to find the best solution.

語　句

strange[streindʒ] 形奇妙な　**operational activity** 作戦行動　**army**[ɑ́:mi] 名陸軍，軍　**bomb**[bɔ́m] 名爆弾 動爆撃する　**airforce**[ɛ́əfɔ:s] 名空軍　**struggle**[strʌ́gl] 動争う，戦う　**invasion**[invéiʒən] 名侵入← invade 侵入する　**air raid** 空襲　**technique**[tekní:k] 名技術，技法　**confront**[kənkrʌ́nt] 動直面する　**situation**[sitjuéiʃən] 名状況　**decision-making** 意志決定　**mathematical**[mæθimǽtikl] 形数学の　**figuratively**[fígjurətiv] 形比喩的に　**formulate**[fɔ́:mjuleit] 動公式化する← formula　公式　**implement**[ímplimənt] 動実行する＝carry out，名道具　**numerous**[njú:mərəs] 形多数より成る，たくさんの　**key**[ki:] 形重要な 名かぎ

●イントロ●　オペレーションズリサーチ（OR）とはよりよい決定をするのを助けるために先進の分析法を適用する手法である．なお，operations と必ず複数形で用いるのは，もともと軍の作戦行動を意味する語だからである．オペレーションズリサーチは複数の作戦が最適な方法で，効率的に実行可能かどうか，リサーチ，つまり検証する，そういう目的で使われ始めた科学であったが，現在では数学や統計学を用いた数理的なモデルで分析することで，最適な手法を検討する経営工学の学問となっている．このオペレーションズリサーチはゲーム理論や金融工学などに広く適用されている．

▶**Though it may sound strange** ➡「それは奇妙に聞こえるかもしれないが」．これは SVO の構文で sound は自動詞，strange はその補語．

▶**it was so severe that England had to struggle** … ➡ so と that は関連している．「それが非常に激しかったのでイギリスは…に努力しなければならなかった」．

▶**mathematical aids where there are two or more alternative courses** ➡関係副詞 where の先行詞は mathematical aids.「二つまたはそれ以上の選択肢があるような数学的支援」．

▶**"Seven unknowns and four equations"** ➡直訳すると「七つの未知数と四つの方程式」．未知数の数と方程式の数とが同じであれば解は厳密に決定されるが，このように未知数の数が多いと解は不定になる．いくつかある解の中から最適のものを選び出すところに OR の出番があるのだといっている．

▶**a model of the system** ➡「システムのモデル」．モデルとは，問題を簡単にするためにシステムから不要なものを捨てて，単純な数式または思考に置き換えたものをいう．

解 説　本文にビルや学校のエレベータやトイレの数を OR で解決する例が述べられているが別の例をあげてみよう．

工場の工具室にテスタが 5 台保管され，20 人の工員が必要に応じてそれを借り出してある検査を実行する．テスタが空いていれば仕事が円滑に進むが，全部出払っていると工員はテスタが戻って来るまで待たなければならない．そこで，テスタを増やせばいいのだが，何台あれば最も適当だろうか．20 台揃えれば工員の仕事は最もうまく進むが，それでは高価なテスタの側にむだが起こる．このような問題を「順番待ちの問題」といって，OR が得意とする部門の一つである．そのほか「割当ての問題」「設備の更新時期の問題」などがある．

問　題

1. 次の（　）の中の語を並べ変えて正しい英文とし，それを日本語に訳せ．
   (1)　OR (originated,　the British army,　from,　the operational activity,　of).
   (2)　OR (applies,　to,　scientific methods and techniques,　determine,　the best solution).
   (3)　OR (has taught,　it,　is,　us,　that,　to make,　very important,　a suitable model).
   (4)　In a sense,　OR (is,　to,　similar,　computer simulation).
2. 本文の内容と合っているものには○，合わないものには×をつけよ．
   (1)　Anti-air raid study was the first step of OR.
   (2)　OR can provide several answers to help make better decisions.
   (3)　OR is only used in the area of manufacturing.
   (4)　OR is an analytical method to help make better decisions.

6 システム工学

System Engineering (SE)

A system usually means a large organization in which interacting and interdependent components are organically combined in order to accomplish one specified purpose. System engineering (SE) is a branch of engineering, which approaches problems by first analyzing the system into factors and then connecting them purposefully through the aid of computers and other devices. Generally, system engineering involves two operations : one is modeling, in which each factor of the system and the standards for performance are clearly described ; and the other optimization, in which many factors are respectively adjusted to obtain the best overall performance.

The idea of system engineering is essential to our daily life. For instance, when cooking meal, we have to consider many factors such as nutrition, economics, skill of a cook, and so on. In a family, these judgements are usually left to the homemaker, but in large cooking facilities, they should be combined systematically to be treated by a computer or by mathematics. One of the most successful instance of SE is the bullet train, or "shinkansen," in Japan, in which a vast number of factors, such as time tables, manpower, and station buildings are integrated into a complicated system to achieve optimum order, safety, and efficiency of transportation.

語　句

organization[ɔ:gənaizéiʃən] 名組織(化) ← organ 機関　**interact**[intərǽkt] 動互いに作用し合う　**interdependent**[intədipéndənt] 形相互に依存する　**component**[kəmpóunent] 形(物の)構成要素である　**approach**[əpróutʃ] 動接近する，近づく　**describe**[diskráib] 動記述する　**optimization**[ɔptimaizéiʃən] 名最適化　**overall**[óuvəɔ:l] 形全体の　**nutrition**[nju:tríʃən] 名栄養　**economics**[i:kənɔ́miks] 名経済学　**homemaker**[hóummeikə] 名主婦　**bullet train** 弾丸列車　**time table** 時刻表　**manpower**[mǽnpauə] 名人力　**transportation**[trænspɔ:téiʃən] 名輸送

●イントロ●　よく話題になっているシステムエンジニアリングは，それぞれ独立した組織や人間が集まって大きな仕事をする際に，いかに科学的に能率よく組織するか，という技術である．

構 文　▶**which approaches problems by first…and then～**➡ first と then とは関連している．「システムエンジニアリングは，まず…を分析し，次に～を統合することによって問題に迫っていく」．

▶**when cooking meal** ➡「食物を調理するときには」．分詞構文で，意味を明確にするために接続詞 when が添えてある．

▶**bullet train, or "shinkansen", in Japan** ➡「日本における弾丸列車，すなわち新幹線は」．bullet train と shinkansen とは同格用法．

解 説　システムを組み立てる専門職であるシステムエンジニアは，システムを構成する各要素に精通し，コンピュータを駆使でき，しかも数学的素養に富んでいなければならないので容易な仕事ではない．システムが膨大になるにつれ，一人の人間の経験や勘ではとても手に負えなくなってきている．そこでこのような新しい職種が生れてきたともいえよう．

問 題

1. 次の（　）の中の語を並べ変えて正しい英文とし，それを日本語に訳せ．
   (1)　System engineering (is,　engineering,　of,　a branch).
   (2)　System engineering (of,　the two operations,　involves,　modeling and optimization).
   (3)　They (combined,　should,　be,　the aid of,　with,　a computer or mathematics).
   (4)　A vast number of factors (integrated,　into,　is,　a complicated system).
2. 本文の内容と合っているものには○，合わないものには×をつけよ．
   (1)　A system means an engineering process to accomplish one specific purpose.

(2)　System engineering is to design and manage complex engineering systems.

(3)　Modeling is an important process in system engineering.

(4)　System engineering is a way to deal with complicated systems.

コラム　イギリス英語とアメリカ英語の違い：スペル

イギリス英語とアメリカ英語では，単語のスペル（綴り）が異なったり，同一のものを異なる単語で表したりすることがある．技術英文を書く際は，どちらかの形式に統一して，イギリス英語とアメリカ英語が混在した文章を作らないことが肝要である．

語尾の re（英）と er（米）：	「中央」	centre（英）	center（米）
語尾の se（英）と ze（米）：	「解析する」	analyse（英）	analyze（米）
語中の our（英）と or（米）：	「色」	colour（英）	color（米）
語中の l（英）と ll（米）：	「移動」	traveling（英）	travelling（米）

7 テクノロジーアセスメント

Technology Assessment (TA)

In the past, technology was always thought to have positive effects on society, and even when its development had a bad influence, such as public nuisance, aftercare countermeasures were only reluctantly implemented. Recently, however, this situation has been changing rapidly. The negative influences likely given by future technical development are now forecast prior to the actual initiation of a project. This negative evaluation is called technology assessment or TA for short. This concept has originated in the United States in 1967 to 1968 to make previous predictions on public nuisance, such as environmental pollution or contamination, and to establish the most suitable counterplan to prevent the negative impact.

In Japan, TA project blossomed in the 70's in such areas as agricultural chemicals on golf links or radioactive waste in atomic plants, and the project still has a strong influence on various fields in Japan. In the Third World, technology impact assessment (TIA) has become a very crucial problem these days, which looks at the inferior impact brought in by high technology projects introduced from developed countries.

語句

assessment[əsésmənt] 名評価，査定← assess 評価する **countermeasure** [káuntəmeʒə] 名対策 **reluctantly**[rilʌ́ktəntli] 副いやいやながら **forecast**[fɔ́:kæst] 動予言する，予想する **initiation**[iniʃiéiʃən] 名開始，創業 **evaluation** [ivæljuéiʃən] 名評価 **prediction**[pridíkʃən] 名予言 **public nuisance** 公害 **pollution**[pəljú:ʃən] 名汚染，公害 **contamination**[kəntæminéiʃən] 名汚染，公害 **counterplan**[káuntəplæn] 名対策 **impact**[ímpækt] 名衝突，影響 **blossom** [blɔ́səm] 動花開く **agriculture**[ǽgrikʌltʃə] 名農業 **radioactive**[réidiouǽktiv] 形放射性の **waste**[weist] 名廃物，むだ **the Third World** 第三世界（発展途上国） **technology impact assessment** 技術影響評価 **developed country** 先進国

●イントロ●　技術評価とは，新技術の急激な適用によって引き起こされる環境や一般社会に対する悪影響を，前もって予測し，対策を立てることをいう．

構文

▶**aftercare countermeasures were only reluctantly implemented.** ➡ aftercare は countermeasures の形容詞と考える．「後始末的対策がやっと実行された」．

▶**The negative influences likely given by future technical development** ➡「将来の技術発展によって引き起されるかもしれないようなマイナスの影響」．

▶**in the 70's in such areas** ➡ 70's は 19 を省略した形で 1970 年代の意．「そのような分野で 1970 年代において」．

▶**…, which looks at the inferior impact…** ➡ which の先行詞はやや離れた technology-impact assessment で，継続用法．「それは…によってもたらされる悪い影響を監督する」．

解説

pollution と contamination とは汚染という意味で同義に用いられることが多いが，contamination は汚染の事実のみを述べるのに対して，pollution はより幅広く経過を含めて記述するときに用いる．公害という意味では public nuisance を用いるのがよい．

問　題

1. 次の（　）の中の語を並べ変えて正しい英文とし，それを日本語に訳せ.

(1) The technology (was, to, have, always, thought, positive effects, on society).

(2) A negative evaluation (called, technology assessment, TA, is, or, for short).

(3) In Japan, the first TA (was, for, agricultural chemicals, on, golf links).

(4) TA is a process (to, evaluate, influence, before, introduction, of a new technique).

2. 本文の内容と合っているものには○，合わないものには×をつけよ.

(1) Contamination by manufacturing activity was positively eliminated.

(2) The concept of TA originated in Japan to predict public nuisance.

(3) TA is an old technique and has no relevance today.

(4) The influence is forecast by TA before actual initiation of a project.

コラム 紛らわしい単語

下記は，似たニュアンスをもちながら明確に異なる意味をもった単語の組合せである．意図に反した誤用を避けるため，よく注意する必要がある．

adapt（～を適応・応用する）	adopt（～を採用する）
alternative（代わりとなる）	alternate（交互の）
effect（(変化・結果など）をもたらす）	affect（～に影響する）
proceed（進める，続行する）	precede（～に先行する）

8 パート

PERT

PERT is an acronym for "program evaluation and review technique," and is a method to plan and control the time schedule of a large-scaled project by using a computer and other aids. This method was initiated in the United States in developing the system for the Polaris missile in the late 50's. The basic features of PERT are：(1) A vast amount of events contained in a project are analyzed and specified into as small one as possible；(2) Small events are laid out in an arrow diagram by which they are subsequently put in order；(3) The relation between them are completely understood；(4) On the diagram, the events or paths thought to make bottlenecks are identified； and (5) These bottlenecks are preferentially controlled to efficiently develop the project.

PERT has a major advantage in that the network of a project is proposed and analyzed in the stage of planning to find impeding bottlenecks. Conventionally, the interdependency and mutual relation among the many events were not necessarily obvious and well defined. However, thanks to the analysis of networks and critical paths in PERT, the points at issue, which may bring about obstacles in smooth progression, are revealed and eliminated in advance.

語　句

review[rivjúː] 名動再検討（する）　**initiate**[iníʃieit] 動着手する　**layout** レイアウトする　**arrow diagram** 矢線図（→解説）　**subsequently**[sʌ́bsikwəntli] 副後で，続いて　**event**[ivént] 名事柄，事件　**bottleneck**[bɔ́tlnek] 名障害，ネック　**identify**[aidéntifai] 動確認する，鑑定する　**preferentially**[prèfərénʃəli] 副優先的に　**network**[nétwəːk] 名網，網状組織　**necessarily**[nésisərili] 副必然的に　**critical**[krítikl] 名重要な，危機の　**point at issue** 問題点　**obstacle**[ɔ́bstəkl] 名障害　**reveal**[rivíːl] 動表す　**eliminate**[ilímineit] 動消去する

●イントロ●　PERTは，日本語で「計画評価および確認の技術」というが，一般には「パート」で通っている．

構文 ▶**…specified into as small one as possible** ➡「…はできるだけ小さい事柄に具体化される」．as…asは同じ程度を表す比較構文．

▶**by which they are subsequently put in order** ➡「矢線図によってそれら（事柄）は次から次へと順序づけられる」．

▶**events or paths thought to make bottlenecks are identified** ➡「あい路を形成していると思われる要素や過程が確認される」．

▶**the interdependency and mutual relation…were not necessarily obvious** ➡「依存関係および相互関係が必ずしも明瞭ではなかった」．not necessarilyは部分否定．notだけでは完全な否定になるが，necessarilyを加えて否定の程度を和らげている．all，every，muchなどを加えても部分否定を作ることができる．Lathes cannot machine every part.「旋盤はすべての部品を加工できるわけではない」．

解説 矢線図（**arrow diagram**）　JIS Z 8121（オペレーションズリサーチ用語）によると，「プロジェクトを達成するのに必要な作業の相互関係を結合点および矢線を用いて図示した手順計画図」と定義されている．一般に物質，エネルギー，情報などの流れる方向を矢印を用いて表した図である．

問　題

1. 次の（　）の中の語を並べ変えて正しい英文とし，それを日本語に訳せ．

（1）PERT (for,　an acronym,　is,　program evaluation and review technique).

（2）PERT (developed,　for,　the Polaris missile program,　was,　in the United States).

（3）PERT (is,　the techniques,　one,　of,　in process planning and management).

（4）PERT (applied,　to shorten,　for,　the schedule,　is,　software development).

2. 本文の内容と合っているものには○，合わないものには×をつけよ．

(1)　PERT is a technique to plan and control the time schedule.

(2)　PERT originated from the operational activity of the British army.

(3)　PERT uses a computer to plan and control the time schedule.

(4)　PERT is almost comparable to OR.

コラム　イギリス英語とアメリカ英語の違い：単語

エレベータ：	lift（英）	elevator（米）
時刻表：	timetable（英）	schedule（米）
ガソリン：	petrol（英）	gas, gasoline（米）
電気コード：	flex（英）	cord（米）
コンセント：	socket, plug（英）	outlet（米）
バッテリー：	accumulator（英）	battery（米）
飛行機：	aero plane（英）	airplane（米）
郵便番号：	postcode（英）	ZIP code（米）

9 非破壊検査

Nondestructive Testing

Nondestructive testing involves the inspection of an object or material to detect faults, determine its properties, or assess quality without breaking down or impairing the quality of the products, and it is considered to be a very useful testing method. Some of the techniques which are employed in medicine, such as radiography（X-ray）and ultrasonics, can be applied in an inspecting shop as well. Typical nondestructive testing methods are as follows：Visual-optical method assesses the product color or surface condition by inspector's observation, and in extreme case, products are checked using a laser beam or complex image-recognition equipment such as an optical-fiber device；Radiographic method uses very powerful radiation as X-ray, gamma ray, or neutron ray to detect flaws hidden inside the material；Ultrasonic method can gather information from ultrasonic waves reflected off any flaws or structural discontinuities present inside the materials；and Liquor penetration method uses dyestuff or fluorescent substances dissolved in oil that are then developed after penetrating into flaws.

語　句

nondestructive[nɔ́ndistrʌ́ktiv] 名非破壊の(non- 否定の意)　**fault**[fɔ:lt] 名欠点　**impair**[impέə] 動悪くする，害する　**medicine**[médisn] 名医学　**radiography**[reidiɔ́grəfi] 名放射線透過法　**ultrasonics**[ʌ́ltrəsɔ́niks] 名超音波法(ultra-超)　**visual-optical method** 目視探傷法　**extreme**[ikstrí:m] 形極度の　**laser beam** レーザ光　**image-recognition** 実体認識　**optical-fiber device** 光ファイバー装置　**radiographic method** 放射線透過検査法　**gamma ray** ガンマ線　**neutron ray** 中性子線　**flaw**[flɔ:] 名欠陥　**ultrasonic method** 超音波探傷法　**reflect**[riflékt] 動反射する　**discontinuity**[dískɑ̀ntənjú:əti] 名不連続← continue 続いている　**liquor penetration method** 浸透探傷法　**dyestuff**[dáistʌf] 名染料← dye 染める　**fluorescent**[fluərésent] 形けい光を発する

V

管理技術

●イントロ●　材料検査を行う際，普通は何らかの傷跡を材料に残すが，本文で述べられている非破壊検査はまったく材料を傷つけない検査法である．

構文 ▶**to detect faults, determine its properties, or assess** … ➡ detect, determine, assess は to にかかる．「欠陥を探したり，その性質を調べたり，また定量試験したりするための」．

▶**in extreme case, products are checked** … ➡「特別な場合には，製品は…で検査される」．

▶**to detect flaws hidden inside**…➡ hidden は hide「隠す」の過去分詞で flaws を修飾している．「物質の内部に隠れている欠陥を探すために」．

▶**dyestuff of fluorescent substances** … **that are then developed** ➡関係代名詞 that の先行詞は少し離れて dyestuff or fluorescent substance.「傷口に侵入し発色するような染料または蛍光物質が用いられる」．

問　題

1. 次の（　）の中の語を並べ変えて正しい英文とし，それを日本語に訳せ．
   (1)　Nondestructive testing (is,　faults,　of products,　to detect, without damaging down).
   (2)　Nondestructive testing (called,　is,　sometimes,　nondestructive inspection).
   (3)　Nondestructive testing (be,　considered,　is,　to,　a very useful testing method).
   (4)　Nondestructive testing (is,　to find,　a method,　inside,　flaws, materials).
2. 本文の内容と合っているものには○，合わないものには×をつけよ．
   (1)　Nondestructive testing is not a reliable testing method.
   (2)　There are various nondestructive testing methods.
   (3)　One of the nondestructive testing methods is the use of ultrasound.
   (4)　X-ray cannot detect flaws hidden inside materials.

○● 章末問題 ●○

1. 空欄を埋めて，日本語に訳せ．

(1) The parts are often made (　　　) factories located hundreds of kilometers apart.

(2) The most important factor is the method of selecting samples (　　　) the lot.

(3) Parts rejected (　　　) inspection should be discarded.

(4) Measured values of characteristic such as dimension, temperature, and concentration are sequentially plotted (　　　) the chart.

(5) OR is defined (　　　) the application of scientific methods and techniques to determine the best solution.

(6) System engineering involves two operations：one is modeling and the (　　　) is optimization.

(7) This concept has originated (　　　) the United States to make previous predictions (　　　) public nuisance.

(8) PERT is a method to plan and control the time schedule of a large-scaled project (　　　) using a computer and other aids.

(9) Nondestructive testing is considered (　　　) be a very useful testing method.

2. 次の文を英語に直せ．

(1) 互換性の概念が大量生産の基礎を築いた．

(2) 抜取り検査とは最も強力で効果的な方法である．

(3) 検査は製品の特性や寸法を正式にチェックすることが必要である．

(4) そのチャートの理論的な背景はシューハートにより提案された．

(5) OR は数学的な助けによって実行される意思決定プロセスである．

(6) それらはコンピュータ処理で系統的に組み合わされなければならない．

(7) TA は日本のさまざまな分野に対して強い影響がある．

(8) パート（PERT）はポラリスミサイル計画のために開発された．

(9) 放射線透過検査法は製品の内部の欠陥を探知する方法である．

VI
文　　　　法
Grammar

母国語でない英語の習得には，文法理解が近道である．文法的な間違いは，ときに読み手の誤解につながる恐れもある．

この章では，初歩的な英文法をまとめて述べている．中学校，高等学校における英文法の復習のつもりで読んでいただきたい．

数式の読み方や，技術英文における心掛けについても簡単に説明している．

1 文型・態

Pattern of Sentence・Voice

1.1 文（Sentence）

下記は，いずれも文である．

- Many machines, such as motors, lathes, and automobiles work with energy. ➡ I−1−1.
- Energy exists in a variety of forms. ➡ I−1−2.

文とはこのように，ある判断を示すもので，判断の主体（この文では many machines と energy）を**主語**といい，判断の内容（work と exists）を示すものを**述語**という．日本語では「アメリカへ行ってきた」のように主語を省くことが多いが，英語では必ず主語と述語を必要とする．

1.2 文字（Letter）

文は語（word）より成り，語は文字で形成される．英語の文字は 26 文字で，それをアルファベット（alphabet）という．これが組み合わされて何十万という語（単語）が作られる．アルファベットは日本の漢字と違ってそれ自身は意味をもたないが，それらは名称（名前）と音(おん)をもっている．A（a）の名称はエイ［éi］であり，音(おん)はときに［éi］であったり［æ］（アとエの中間的な音）であったりする．many の "a" は［é］であり，machine の "a" は［ə］（曖昧母音とも呼ばれ，口の中をリラックスさせて自然に出る音）である．このように一つの文字がいろいろな音(おん)を持っていることは英語の特徴であり，発音を難しくしている原因の一つである．

1.3 語（Word）

文字がいくつか集まってつくられるのが**語**である．e.n.e.r.g.y の文字一つひとつは意味をもたないが，ある順序で並べられると energy「エネルギー」（発音は［énədʒi］エナジィ）という意味をもつ単語ができる．文字の並べ方の決まりをつづり（spelling：スペリング）という．

1.4 文法（Grammer）

単語を連ねて文をつくる際の決まりを**文法**という．「英文法など習う必要はない．私達は日本語の文法を知らなくても正しい日本語を話したり，書いたりするではないか」といわれることもあるが，これは誤りである．私達が日本語を正しく使えるのは，膨大な日本語の文法を脳のメモリーの中に意識せずに蓄えているからである．ところが英語の場合は，そのような英文法のソフトが私達の脳のメモリーの中にまったく存在していない．たとえば，幼少から英語のシャワーを浴びていれば，自然と馴染みのある言葉の法則（つまり英文法）が体得できるかもしれない．しかし，論理的にものごとを考えられるような成人になってからは，規則立てられた言葉の法則を学ぶほうが効率的である．したがって，英文法という形で意識的に記憶していくことが，正しい英文の読み書きへの近道である．面倒でも**英文法の基礎を理解し記憶する**必要がある．

英語には日本語の**助詞**（てにをは）に相当する言葉がない．ある語が主語か述語か，または**目的語**か区別するには，その置かれた位置によって判断する．たとえば，

Makoto loves Akemi.

とあればMakotoが主語，lovesが述語，Akemiが目的語であると判断される．これが，

Akemi loves Makoto.

となれば立場が反対になる．これと対照的に，日本語ではそのままの語順で，助詞を変えることにより「マコトはアケミを愛している」と「マコトをアケミは愛している」のように立場の反転した文をつくることができる．このように考えると，英文法とは，文中の英単語の正しい並べ方の研究の学問であるともいえる．

1.5 自動詞と他動詞

文中の述語に用いられる動詞はその働きによって，次の二つに分けられる．動作を行うときに相手の要らない**自動詞**（intransitive verb）と，相手・動作対象が必要である**他動詞**（transitive verb）とがある．たとえば，A dog runs.「犬が走る」やI walked.「私は散歩した」では，「走ったり」「散歩する」のに相手が要らない．したがって，動詞runとwalkは自動詞である．

一方，I have a nice friend.「私はよい友人をもっている」では，have「もつ」のに相手（この場合 friend）が必要である．このような動詞を他動詞といい，動作の対象を**目的語**（object）という．目的語には名詞（相当の句や節）が用いられる．

他動詞は一般に目的語を伴うため，受動態の構文に替えることができる．数ある動詞の多くは自動詞と他動詞に区別することができるが，同一の単語が両方の用法で用いられる動詞もある．たとえば，

In a battery, chemical energy is transformed into electrical energy. ➡ I −1−4.
In a battery, chemical energy transforms into electrical energy.
の2文の構文は異なる（上は他動詞で受動態の文，下は自動詞の文である）が，ほぼ同じ意味を述べている．あえて違いを挙げると，上の文では動作主（by に続くもの）は省略されているが，be transformed「変換させた」“何か”があることを意味する．一方，自動詞の場合は，“自然と自ら”エネルギーが変換したイメージを伝えている．

This is.「これは＿である」，または　I become.「私は＿になる」
の文で is も become も自動詞であるが，これだけでは文は不完全である．これらに語を補って，たとえば
This is beautiful. または I become an engineer.
とすれば「これは美しい」，「私は技術者になる」と完全な文になる．この beautiful や engineer を**補語**（complement：これらは主語である This や I に相当するものなので**主格補語**）といい，このように補語を必要とする動詞を**不完全自動詞**という．

He makes her.「彼は彼女を＿にする」
の文も不完全である．
He makes her happy.
のように happy を補えば「彼は彼女を幸福にする」と完全な文になる．make は her を目的語とする他動詞であるが，目的語だけでは不十分で，その後に目的語を説明する**補語**（**目的格補語**）を必要とする．このような他動詞を**不完全他動詞**という．

不完全他動詞のうちで make のように「人に＿させる」の意味をもつものを特に**使役動詞**（causative verb）という．**make** のほか **let**，**allow**，**permit**，

leave などがよく用いられる.

The cams allow the intake valves to open.「カムは吸気弁を開くようにする」
では to open が目的格補語になっている.

give（与える）は他動詞であるが，この他動詞は二つの目的語，つまり「与える相手（**間接目的語**）」と「与える品物（**直接目的語**）」を必要とする．このような他動詞を**授与動詞**（dative verb）という.

I gave him a book.「私は彼に本を与えた」
では him が間接目的語，a book が直接目的語である．**ask**（尋ねる），**send**（送る），**show**（見せる），**teach**（教える）などが主な授与動詞である.

1.6 能動態（Active voice）と受動態（Passive voice）

他動詞を使った文では，動作の及ぶ相手，目的語が必要である．その際に動作をする主体が主語になっている文を**能動態**といい，逆に動作が及ぶ相手（能動態における目的語）が主語になっている文を**受動態**（受け身）という.

Sam loves Nancy.「サムはナンシーを愛している」
では主体 Sam が主語になっているので能動態である．愛される Nancy を主語にして，
Nancy is loved by Sam.「ナンシーはサムに愛されている」
とすれば受動態の文になる．どちらも同じ内容を表現しているが，技術文書では比較的受動態の文を用いることが多い.

受動態の文は「be 動詞＋過去分詞」の形となる．その後に by（ときに with など）を用いて動作の主体を導くことがある.

1.7 五文型（Five patterns of sentences）

前述のように，動詞には自動詞と他動詞に大別されるが，さらに区別すると完全自動詞，不完全自動詞，完全他動詞，不完全他動詞，授与動詞の 5 種類がある．これに伴って，文にもそれぞれの動詞を用いた五つの型がある．これを五文型といい，次の表のようになる.

動詞の種類			文型
自動詞	完全自動詞	(complement int. verb)	S+V
	不完全自動詞	(incomplement int. verb)	S+V+C
他動詞	完全他動詞	(complement tra. verb)	S+V+O
	不完全他動詞	(incomplement tra. verb)	S+V+O+C
	授与動詞	(dative verb)	S+V+IO+DO

S は主語（Subject），V は動詞（Verb），C は補語（Complement），O は目的語（Object），IO は間接目的語（Indirect Object），DO は直接目的語（Direct Object）を表す．

すべての英文は五文型のいずれかにあてはまる．実際の英文が長いのは，主語，述語，目的語，補語にいろいろな飾りの語がつくからである．これを**修飾語**という．たとえば，

In the same way that water flows through a pipe, electricity flows along a solid wire. ➡ I -6-7.

は長い文であるが，基本は electricity flows.「電気は流れる」の S + V 構文であって，それ以外の語句は修飾のためにつけられているに過ぎない．また，

Whether the drill is large or small, carbon or high-speed steel, held in a chuck or otherwise, and used with a jig or without, the principles of its operation are nearly the same. ➡ Ⅲ-4

も，Whether の箇所が長いだけで，the principles ... are ... the same（原理は同じである）の S + V + C 構文である．より丁寧に解釈すれば，

Whether S + V + “C or C”，“C or C”，“C or C”，and “C or C”，S + V + C.

という構文になっているが，結局はいずれの文も五文型のいずれかにあてはまる．

1.8 形式主語としての it

It is important to obtain the knowledge of and skill with a drilling machine and tools for the practice in a machine shop. ➡ Ⅲ-4

の It は，to 以下の内容を表すための形式主語として用いられている．「ボール盤と刃物についての知識をもち，熟練を積むこと（→この内容が It）が重要である」となる．真の主語が that 以下で導かれる場合もある．

It has been found that speeds can be up five times.「スピードが5倍にもなるということ（→この内容がit）が見出された」

形式主語itを用いる理由は，文の頭を軽くして文を読みやすくするためである.

1.9 命令文（Imperative sentence）

命令文とは，「__せよ」という日本文に相当する英文の形で，(1) 主語を省き，(2) 動詞の原形を文頭に置く形式で述べられる．日常の会話でも用いられるほか（例：Come here.「ここへ来なさい」），技術文書でもしばしば用いられる．特に，操作法や注意書などは命令形で書くのがよい.

Use only screwdrivers that fit the screw slots correctly.「ねじの頭に正しく合ったねじ回しを使いなさい」

「__するな」という命令文をつくるには，Don't あるいは Never を文頭へ置く.

Don't change spindle speed until the lathe comes to a dead stop.「旋盤が完全に止まるまでは主軸のスピードを変えてはいけない」

なお，never のほうが don't よりも強い禁止を意味する.

2 名詞・冠詞

Noun・Article

2.1 名詞（Noun）

“もの”の名称を名詞といい，次のように分類される．

- **固有名詞** 一つしか存在しないものの名称（例：Pascal　パスカル氏，Tokyo　東京）
- **同種類のものに共通に用いられる名称**
  - **可算名詞**（数えることのできるものの名称）
    - **普通名詞** 同種類の事物に共通に用いる名称（例：desk　机，book　本，lamp　ランプ）
    - **集合名詞** 集合体の名称（例：people　人々）
  - **不可算名詞**（数えることのできないものの名称）
    - **物質名詞** 材料・物質の名称（例：silver　銀，wood　木材，steam　水蒸気）
    - **抽象名詞** 物質や動作の名称（例：work　仕事，beauty　美しさ，diligence　勤勉）

2.2 単数と複数（Single and plural）

英語では単数（1個）と複数（2個以上）の区別が厳密であるから，複数の事物を表す名詞には語尾に必ずsをつける．たとえば，

dog → dogs，book → books

不規則に変化するものとしては，

child → children，foot → feet，man → men などがある．

複数形が単数形と異なる意味をもつ名詞がある．たとえば，arm（腕）とarms（武器），good（善・利益）とgoods（貨物・品物），work（仕事）とworks（工場）などがある．

2.3 所有格のs（“s” of possessive case）

foreman's one touch「職長のワンタッチ」➡Ⅳ-6

の -'s（アポストロフィs）は所有の意を表す形容詞（**所有格**）を作る．このよう

に -'s がつけられるのは所有者が人間のときだけで，人間以外には the legs of the table「テーブルの脚」のように of を用いる．例外的に，Earth's surface とすることができるのは，地球，太陽などの絶対的存在を擬人的に扱っているためである．複数名詞の所有格でも -'s をつけるが，語尾が -s である場合は -s's とせず，-s' と記す．

workers' market　　children's books　　Ken's family　　Toms' effect

2.4 頭字語（Acronym）と短縮語（Abbreviation）

複数の語から成る複合語を使いやすいように省略する方法で，各語の頭文字を述べたものをいう．Computer-Aided Design を CAD とし「キャド」と読むのは**頭字語**である．UNESCO（United Nations Educational, Scientific, and Cultural Organization）は「ユネスコ」と読み，やはり頭字語である．

一方，IBM（lnternational Business Machine Corp.）は「アイビーエム」と1文字ずつ読むので**短縮語**であるが，この区別は厳密なものではない．紙面や時間の節約のために，最近は頭字語や短縮語がますます使われる傾向にある．

2.5 複合語（Compound word）

heat-resisting metal（耐熱性金属）➡ Ⅲ-10 では heat と resisting とが一つの語を作って metal を修飾している．このように，2語以上が結ばれて一つの名詞や形容詞を作る場合にそれを**複合語**といい，その間をハイフン（-）で結ぶ．

つまり，先の例では，heat が resisting を修飾し「熱に抵抗する＝耐熱性の」という意味を成して metal を修飾する．ハイフンがないと heat と resisting が共に metal を修飾するため「熱い抵抗金属」という意味になってしまう．複合語は使われているうちにハイフンが取れて，完全な1語になってしまう場合もある．

●その他の例

temperature-dependent change　　温度依存変化
time-averaged value　　時間平均値
alcohol-free products　　ノンアルコール製品

また，physically measured value「物理的に測定された量」，commonly adopted sensor「一般に採用されているセンサ」では，いずれも初めの2語が終わりの語を修飾している．上述のようにハイフンを入れてないのは最初の語が

-lyによる副詞（動詞や形容詞を修飾し，名詞は修飾しない）であり，初めの2語がひとまとまりで最後の語（名詞）を修飾していることが明らかなためである．

2.5 冠詞（Article）

形容詞の一種で，不定冠詞（a/an）と定冠詞（the）の二つだけであるが，日本語にはそれに相当する語がないので正しい使い方は難しい．

a/anは「一つ」を意味している．

I have a pen.「私は1本のペンを持っている．」

のように，一般的に可算名詞の単数にはa/an（またはthe）が必ずつく．theはその名詞が何を指すかが話者および聞き手にわかっている場合に用いられ，「その」と訳す場合が多い．

I have a cat, and the cat catches rats usually.「私は1匹の猫を飼っている．その猫はいつも鼠をつかまえる．」

文章に名詞が初めて現れる場合には，一般にaで修飾し，それ以後はこの名詞はtheで修飾されると考えてよい．

冠詞に不慣れな日本人の英語では，冠詞を付け忘れがちだが，これにより文の意味が異なってしまうこともあるので注意する．

- I met an engineer and manager.「技師兼管理者の方と会った」
- I met an engineer and <u>a</u> manager.「技師と管理者に会った」

「一つの」を強調して伝えたい場合には，a/anの代わりにoneを用いるとよい．

a/anが省略された可算名詞は，不可算名詞として性質・材料・手段などの概念的なイメージで意味を捉えられることがある．たとえば，

- Hold this with a screw.「（1本の）ネジで止めよ」
- Hold this with one screw.「ネジ1本で止めよ」
- Hold this by screw.「ネジ（ネジ止めという手段）で止めよ」

上記とは逆に，不可算名詞でも，ある単位をイメージしながら可算名詞として扱うこともある．たとえば，temperatureとmetalはそれぞれ「温度“という概念”」や「金属“という材料”」を指す不可算名詞の代表例であるが，

a temperature of 100℃「100度の温度」

two metals「2種類の金属」

のように，“値”や“種類”という単位で捉えて可算名詞扱いにできる．

3 句・品詞

Phrase・Word Class

3.1 句（Phrase）

句や節は文の一部であって，構造的にあるまとまりをもったものをいう．そのなかでも**句**とは，いくつかの語が集まって一つの品詞と同じ働きをしているものをいう．品詞には八つの種類があり，句にも名詞句，代名詞句，形容詞句，動詞句，副詞句，句前置詞，接続詞句，および間投詞句の８種類がある．

fuel for jet airplanes では，for jet airplanes は fuel を形容している形容詞句であり，In addition to fuel では，in addition to は接続詞句である．

3.2 八品詞（Eight parts of speech）

英語の単語は文の中における働きによって，次の8種類のどれかに分類される．

① **名詞**（noun）➡ものの名称を表す．

② **代名詞**（pronoun）➡名詞の代わりに用いられる．

③ **形容詞**（adjective）➡名詞の前に置かれ，名詞を修飾する．

④ **動詞**（verb）➡動作や状態を表す．文の述部に使われる動詞を述部動詞という．

⑤ **副詞**（adverb）➡動詞，形容詞，副詞などを修飾（説明）する．

⑥ **前置詞**（preposition）➡名詞の前に置かれ形容詞句や副詞句を作る．

⑦ **接続詞**（conjunction）➡語，句，節，文などを結びつける．

⑧ **間投詞**（interjection）➡「ああ」，「まあ」などの不意に発する発声，感嘆．

ほとんどすべての語は，いくつかの品詞を兼ねている．たとえば This is a dog. の this は代名詞であるが，This dog is black. の this は形容詞として用いられている．

3.3 代名詞（Pronoun）

If multiples or subdivisions of this unit are needed, we can obtain them merely by attaching a prefix to the base unit. ➡ I -9-4.

の文中で，we（私達は）や them（それらを）は何を指しているのだろうか．文脈から判断して，we は一般の人，them は multiples や subdivisions であることがわかる．

このように，ある名詞の代わりに用いられるものを**代名詞**という．代名詞は同一の名称の繰返しを避け，文章を簡潔にするために用いられる．英語は言語の性格上，代名詞の使用が日本語よりも多い．代名詞を分類すると次のようになる．

① **人称代名詞** ➡ 人間について用いられる：I，he，you，we

② **指示代名詞** ➡ 物事を指し示すのに用いられる：that，this，it

③ **不定代名詞** ➡ 漠然と一般の人や物を指すのに用いる：one，some

④ **疑問代名詞** ➡ 疑問文で用いられる：what，who

⑤ **関係代名詞** ➡ 接続詞の働きを兼ねる代名詞：which，who，that

3.4 形容詞（Adjective）

形容詞には physical quantity「物理的な量」の physical のように名詞を直接に修飾する場合と，It is necessary to...「それ（to...以下）は必要である」の necessary のように不完全自動詞の補語として用いられる場合とがある．前者を形容詞の**付加用法**（**限定用法**），後者を**叙述用法**という．形容詞による比較は原級，比較級，最上級による三つの比較がある．

① **原級** ➡ 通常，形容詞の原級の前後に as...as を置く．
He is as tall as his father.「彼は父と同じくらい背が高い」

② **比較級** ➡ 形容詞の比較級を用いて二つのものを比較する．比較級を作るには「原形＋er」，2 音節以上の長い形容詞のときは「more＋原形」とする．比較されるものの前に than「…よりも」を置く．
The temperature is higher than 200℃ .「温度は 200℃よりも高い」

③ **最上級** ➡ 三つ以上のものの間で程度が最も著しいことを示す．最上級を作るには「the 原形＋est」，または「the most＋原形」とする．比較するグループの前に of「…の中で」を置く．
The temperature of today is the highest of these ten days.
「今日の温度はこの 10 日間で最も高い」

3.5 副詞（Adverb）

形容詞と同様に，あるものを修飾することばであるが，形容詞は名詞を修飾するのに対し，副詞は動詞，形容詞，副詞などを修飾する．

Each of 5.02 or 0.0502 clearly has three significant figures. ➡ I −11−4

での clearly は has という動詞を修飾している副詞である．副詞の置かれる位置は比較的自由であるが，一般には修飾する語の近くに置かれる．動詞を修飾する時はこのように動詞の前に置く．

Obviously, 5.020×10^3 is a more precise expression than 5.02×10^3. ➡ I −11−8

での obviously も副詞であるが，これは文全体を修飾しているために文頭に置かれている．これを**文副詞**という．副詞にも比較級，最上級の用法があるが，最上級に the は不要である．

He studied hardest of all students.「彼は全生徒の中で最もよく勉強した」

3.6 前置詞（Preposition）

前置詞はその後に置かれる名詞と一緒になって，形容詞句または副詞句を作る．これらの句の多くは修飾する語の後に来る．前置詞の数は少ないので容易に覚えられるが，さまざまな名詞と結んで意味の微妙に異なった無数の句を作る．

たとえば，various properties of substances では，properties は various と of substances（形容詞句）の二つに修飾されて「物質のいろいろな性質」の意となる．

It decreases by the same factor.

では，前置詞 by は the same factor と結んで「同じ割合で」という意味の副詞句を作っている．前置詞の後ろに置かれた名詞は，前置詞の目的語という．これは他動詞の目的語とは異なるので注意が必要である．

3.7 接続詞（Conjunction）

scientific and engineering の and や，multiples or subdivisions の or は接続詞で，二つの語，節，や文を結びつける．and や or のように両者をほぼ対等に結びつけるものを**等位接続詞（Co-conjunction）**という．

これに対して，If multiples or subdivisions of this unit are needed, we can... の if のように従属的に結びつけるものを**従属接属詞（Sub-conjunction）**という．この文では後半の節が主節になる．

3.8 準動詞（Verbals）

動詞は文の述部動詞以外に，形を変えて，名詞，形容詞，あるいは副詞として用いられる．これらを**準動詞**という．準動詞は次のとおりである．

種　類	形　態	用　法	例
不定詞	to＋原形	名詞，形容詞，副詞	to write，to wish
現在分詞	原形＋ing	形容詞	writing，wishing
過去分詞	原形＋ed，その他	形容詞	written，wished
動名詞	原形＋ing	名　詞	writing，wishing

3.9 不定詞（Infinitive）

不定詞は準動詞の一種で，「to＋動詞の原形」の形で表されるが，to のないものもある．次の三つの働きがある．

●名詞不定詞

To speak English is not easy.「英語を話すことは容易ではない」

では，to speak は名詞の働きをしており，これを名詞不定詞という．形式主語 it を使って，It is not easy to speak English. とすることが多い．

I like to play football.「私はフットボールをするのが好きだ」

では，to play が like の目的語となっている．to play の後に前置詞を介さずに football を置いているのは，to play が名詞と同時に動詞としての働きを残しているためである．

●形容詞不定詞

I have friends to help me.「私は助けてくれる友人をもっている」

では，to help me は friends を形容しているので形容詞不定詞である．このとき，friends と help は主語と動詞の関係にある．ここで，help の目的語 me が削除された場合，friends to help で「助けるべき友人」とも解釈でき，friends と help が目的語と動詞の関係になることに注意する．

●副詞不定詞

I went to meet my father.「私は父に会いに行った」

では，to meet が went という動詞を修飾しているので副詞不定詞である．

to pull a metal test piece「金属試験片を引っ張るのに必要な」➡Ⅱ−7

の to pull は副詞不定詞である.

3.10 動名詞（Gerund）

動名詞は準動詞の一種である. 名前が示すように名詞であるから, 主語, 目的語, または補語となりうる. しかし普通の名詞と異なるところは, 動詞としての性質を残しているので,
①副詞で修飾される（例：completely mixing of metals「金属の完全な混合」）
②前置詞を介さずに直接に目的語を取ることができる.

動名詞は動詞の現在分詞とまったく同じ形をしているので, その用法を混同しないことが大切である. たとえば,
My daughter is the dancing girl with a dancing teacher.「踊りの先生と踊っている少女が私の娘です」
では dancing teacher「踊りの先生」では dancing は動名詞として, dancing girl「踊っている少女」では現在分詞として使われている.

3.11 分詞（participle）

分詞には, 現在分詞（Present participle）と過去分詞（Past participle）による修飾がある. 現在分詞は進行形を作り, 過去分詞は完了時制や受動態を作るが, そのような本動詞としての働き以外に, 準動詞として形容詞の働きもする. その働きは,

① 現在分詞の形で「…する」,「…している」と能動的に修飾する（たとえば roughing reamer「荒削り用リーマ」, revolving drill「回転しているドリル」）.
② 過去分詞の形で 「…された」,「…されている」と受動的に修飾する. the heat generated by electric current「電流によって生じた熱」➡ Ⅲ-8 の generated by は the heat を修飾している.

一般の形容詞は名詞の前に置かれるが, 分詞の場合は名詞の後に置いてもよい. 特に, 分詞に目的語や副詞がついている場合は必ず後ろに置かれる. たとえば size specified on the blue print「図面上に決められた寸法」などのようになる.

4 節

Clause

4.1 節（Clause）

節とは，文の一部であるが，その中に主語と述語とをもっている．He went out, and she came in. という文中で and をはさむ二つの部分は，それぞれ主語と述語をもっているので節である．

4.2 関係代名詞（Relative pronoun）

「私がもっている本」を英語で表す場合は，「私がもっている」という形容詞句を「本」の後に置いて the book which I have とする．この which を関係代名詞といい，the book という節と I have という節とを結びつける接続詞の働きと，the book を表す代名詞の働きとを兼ねている．book を which の**先行詞**という．関係代名詞は先行詞の種類によって主に次の三つが用いられる．

① who（先行詞が人間の場合）

the boy who knows you「君を知っている少年」➡ 関係代名詞 who が knows の主語となっている．

② which（先行詞が人間以外のものである場合）

the desk which has four legs「鉢本の脚のある机」➡ 関係代名詞 which が has の主語となっている．

③ that（あらゆるものに用いられる）

the pen that you gave me「君が私にくれたペン」➡ 関係代名詞 that が gave の間接目的語となっている．

all edges that can be seen「見ることができるすべての外形」での that や，the edge of the object from which measurements are to be made「そこから測定が行われる物体のへり」での which が関係代名詞となっている．

4.3 関係副詞（Relative adverb）

関係代名詞はそれに後続する節の中で主語か目的語になっているのに対して，関係副詞は副詞の働きをしている．この両者をまとめて**関係詞**（relatives）という．a house where I live「私が住んでいる家」では，「私の住んでいる」状態が副詞的に where で修飾されている．which を使って a house in which I live としても同じである．主な関係副詞は次のとおりである．

① the day when I went out「私が出発した日」➡ 出発を when 以下が修飾．

② the town where I was born「私が生れた町」➡ 生れを where 以下が修飾．

③ the reason why I study「私が勉強する理由」➡ 勉強を why 以下が修飾．

④ how I study「私が勉強するやり方」➡ how 自身が先行詞に含まれる．

この場合に，day，town，reason を関係副詞 when，where，why の先行詞という．一般に，how の場合には先行詞を置かない．

4.4 関係詞の限定用法（Restrictive use）と継続用法（Continuative use）

関係詞（関係代名詞と関係副詞）の前にカンマ（,）のない場合を関係詞の限定用法といい，後ろから訳していくことが多い．これに対して（,）のある場合を継続用法といい，前から順次訳していくことが多い．

① We use a gauge block which [that] is a hardened steel block.「私達は，焼き入れした鋼で造られたブロックゲージを使う」➡ **限定用法**

② We use a gauge block, which is a hardened steel block.「私達はブロックゲージを使う．そして，それは焼き入れした鋼で造られる」➡ **継続用法**

①の限定用法では，暗に他の工法・材質で造られたブロックゲージもあることを，ほのめかしている（実際にあるかどうかは確定していない）．

②では，使うブロックゲージは一つであることを伝え，そのブロックゲージの追加説明を後半で行っている．2 文に分けても差し支えない内容である．

Gears are the device which transforms power from one to another.

上文の which も限定用法なので，「歯車は一方から他方へと力を伝える装置である」と訳す．同様に，関係副詞でもこの二つの用法がある．

4.5 分詞構文（Participial construction）

分詞構文は，準動詞である現在分詞（-ing 形）が，接続詞と動詞とを兼ねて副詞節を導くものである．

The density of carbon fiber is usually as low as 1.8 g/cm^3, giving it a very large specific strength ... ➡ Ⅱ−11

giving 以下が分詞構文で，これを書き直せば，

The density of carbon fiber is usually as low as 1.8 g/cm^3, and therefore the low density gives it a very large specific strength ...

となる．意味は「したがって，密度が小さいことはそれに極めて大きい比強度を与える」となる．この場合，本文と分詞構文をつなぐ接続詞として therefore が適切であるが，普通は次の四つの接続詞のいずれかに当てはまる．

① **時**（when）➡ Walking along the street, I came across a friend.「私は街を歩いているときに友人に出会った」（＝When I walked, ...）

② **原因・理由**（as）➡ Living in a city, I have many friends.「私は都会に住んでいるので多くの友人をもっている」（＝As I live in...）

③ **条件**（if）➡ Going on this way, you can find the house.「この道をどんどん行けば，君はその家を見つけられます」（＝If you go on this way, ...）

④ **譲歩**（though）➡ Thinking about your right, I can't forgive you.「君の権利を考えても私は君を許せない」（＝Though I think about your right, ...）

4.6 分詞構文の変形

分詞句の主節に対する関係をわかりやすくするために，when，while，as などの接続詞を分詞の頭に添えることがある．

Walking along the street, I came across a friend.

という文に接続詞 when をつけて When walking along the street, …とすれば一層わかりやすい分詞構文となる．

分詞句の意味上の主語は，主節の主語と同じであることが原則であるが，異なる場合もある．その時は分詞句の主語を書き添える必要がある．

The weather being fine, I visited my uncle's house.「天気がよかったので伯父さんの家を訪問した」

分詞句の主語は weather，主節の主語は I である．

次のような成句は**不人称分詞構文**（impersonal participial construction）といわれ，分詞句の意味上の主語は習慣的に省かれる．

Generally speaking, Japanese are workaholic.「一般的にいえば日本人は働き中毒だ」

5 時制・助動詞

Tense・Auxiliary Verb

5.1 過去・現在・未来(Past, present, and future)

動詞の過去形を作るには「原形+ed」が原則であるが，be → was/were, go → went, have → had のような不規則動詞も多い．また，未来を表す（動詞に「未来形」は無い）には shall や will のような助動詞，または be going to の句を動詞の前に置く．

- This machine worked one year ago.「この機械は1年前には動いた」
- This machine works every day.「この機械は毎日動いている」
- This machine will work tomorrow morning.「この機械は明朝動くだろう」

5.2 進行形時制（Progressive tense）

習慣的に行われている動作は，動詞の現在形で表されるが，これは時間的にかなりの幅をもった表現で，必ずしも「現在…している」を意味しない．「現在…している」を表すには現在進行形を用いる．形は「be動詞＋現在分詞」である．

We are living in the machine age.「私達は現在機械の時代に生きている」
の文は現在進行形の用法である．

過去のある時点で動作が進行中の場合は過去進行形を用い，未来のある時点で進行中を表現するには未来進行形を用いる．

- She was playing the piano when I entered the room.「私がその部屋に入った時に彼女はピアノを弾いていた」➡ 過去進行形.
- I shall be waiting for you at eight o'clock tomorrow morning.「明朝8時に君を待っています」➡ 未来進行形.

5.3 完了時制（perfect tense）

過去の行動が現在まで影響を及ぼしているときに用いられる時制で，「have (has) ＋動詞の過去分詞」の形で表される．実際には次の三つの情況のどれかに当てはまる．

① **完了・結果**（今…してしまった）➡ I have just written the letter.「私はち

ょうど今手紙を書きあげた」

完了・結果の時制には already や just を伴うことが多い.

② **経験**（今までに…をしたことがある）➡ Have you ever been in America?「君はアメリカにいたことがありますか」

③ **継続**（現在も引き続いて…している）➡ I have lived here for five years.「私は5年間ここに住んでいて今も住んでいる」

実際文中に現在完了があったときに，上記のいずれの場合に当てはまるか判断しにくいこともあるが，おおむね文脈によって判断できる.

過去形と現在完了形の違いは，過去のその動作や出来事が現在に（影響が）及んでいるか否かである．過去時制はあくまで過去の出来事を述べているだけであり，現在の状況とは関係がない．一方，現在完了形は現在にまで影響が残っている状態，つまり現時点での「完了・結果・経験・継続」状況を示すために用いられる.

- I stopped smoking last year.「昨年，禁煙した（が，今年は不明）」
- I've stopped smoking for two years.「2年間（今も），禁煙中」

過去完了時制は過去のある時点で既に完了している動作を記述する時に用いる．現在完了時制と同様に，完了結果，経験，継続を表すことができる．その形は「had＋過去分詞」となる.

- I had just finished my homework when I got a call from him yesterday.
  「昨日彼から電話を受けたとき，ちょうど宿題を終えたところだった」

また，未来の一時点で完了するであろう動作を記述するには，**未来完了時制**を用いる．その形は「will，shall have＋過去分詞」となる.

5.4 仮定法現在（subjunctive present）

文章を書いたり話したりするとき，その人の気持ちによって文章のスタイルが違ってくる．この表現の違いを**法**（mood）といい，さまざまな分類方法があるが，次のように分類することが多い.

① **直説法**（indicative mood）➡ ほとんどの文はこの法で書かれる.

② **命令法**（imperative mood）

③ **仮定法**（subjunctive mood）➡ 名前が示すように，事実ではなく「もしも…であれば」と仮定したり，「…したいものだ」と願望するときに用いられ

る法をいう．

ここで，以下の文をみてみよう．

If the size ratio of two gears is 1 to 2, the speed ratio would become 2 to 1.

「二つの歯車の大きさの比が 1：2 であるならば速さの比は 2：1 になるだろう」

この文では，前半の節を仮定節といい，if を使うことが多いので if clause ともいう．この場合，後半の節は助動詞 would を使って「…だろう」と推定を示している．

なお，仮定法現在以外の仮定法は，技術文書ではそれほど必要性は高くない．

5.5 助動詞（Auxiliary verb）

助動詞は，動詞（本動詞）の前に置かれて動詞の意味に変化を与えるものである．助動詞のつかない，現在形・過去形・完了形の文章は，客観的事実や 100 %の確率（疑いの余地のない）出来事などを伝えるものである．つまり，書き手の主観・意思・予測・許容などが含まれていない．助動詞は，書き手のそのような「気持ち」をつけ加える役割をもっている．

客観的事実を扱うことの多い技術文書では，不必要に助動詞を用いるべきではないが，書き手（ときには設計者や製造者）の意図・注意（事故の可能性や予測）を伝える必要がある際には，適当に助動詞を扱うべきである．主な助動詞を以下に示す．

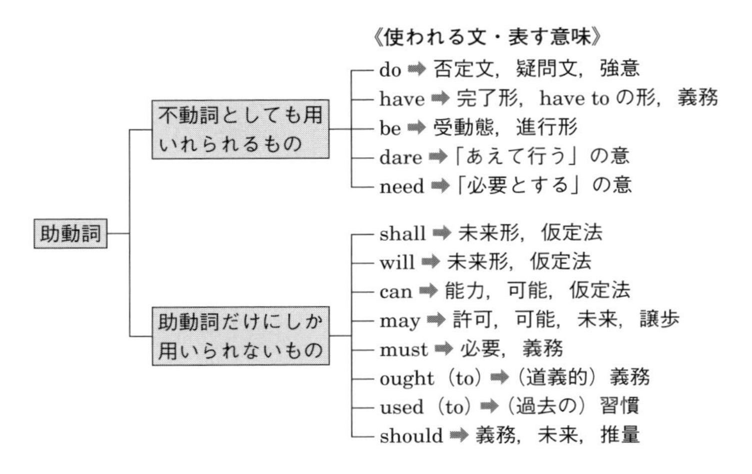

助動詞には似た意味をもつ単語が複数存在するが，厳密には異なることを認識しておくとよい．たとえば，予測や（100 %でない）確信を伝えるために用いられる助動詞「～であろう，～かもしれない」として will，can，may，must，should がある．しかし，これらは確信度合いやニュアンスについて差があり，それを意識して使い分けるとよい．

must ➡（確信度：とても強い．必然性に基づく確信．）
will ➡（確信度：強い．信じて疑わないほどの確信．）
should ➡（確信度：中程度．必然性に基づく確信．）
can ➡（確信度：やや弱い．少しでも可能性がある．）
may ➡（確信度：弱い．否定はできない程度の確信．）

義務を課す助動詞「～すべきである」には，shall，must，should がある．特に，shall は堅いイメージがあるため，口語ではあまり用いられない．助動詞ではないが，had better は should より，やや強い印象で命令的になる．

shall ➡（拘束力：強い．規則・法律・契約などに基づく義務．）
must ➡（拘束力：強い．必然性・必要性からくる義務．）
should ➡（拘束力：弱い．道徳的・状況から判断して生じる義務．）

6 数式などの読み方

How to Read Mathematical Formula

6.1 数詞（Numerals）

数の表し方には，3 種類ある．

① **基数詞** ➡「1 個，2 個…」「一つ，二つ…」などの数える語．形容詞や名詞として扱う．

② **序数詞** ➡「1 番，2 番…」「第 1，第 2…」などの順序を表す語．「第～番目」と特定のものを指すため，通常 the を伴う．

③ **倍数詞** ➡「2 倍，3 倍…」「1 回，2 回…」などの倍数，回数，度数を表す語．一部例外を除き，基数詞を用いた「～ times (as ... as)」で表す．

(1) 基数詞のルール

● 21 以降では，十の位と一の位の語句の間にハイフン（ - ）をつける．

one（1），two（2），…，eleven（11），twelve（12），thirteen（13），…，twenty（20），twenty-one（21），…，thirty-one（31），…

● 英文中において，文頭や 1 桁と 2 桁の数は文字で表すことが一般的である．

（×） 35 divided by 5 is 7.

（○） Thirty-five divided by five is seven.

● 大きな数字は 3 桁ごとに，カンマまたはスペースで区切る．

● 101 以上の数では，百の位と十の位の間に and を入れる（英用法）．

7,329,911,455 → seven billion, three hundred twenty-nine million, nine hundred eleven thousand, four hundred (and) fifty-five

● 正確性を重視する技術英語では，できる限り billion の利用は避け，7,349 million などと記すか，数字を併記することが賢明である．米用法では billion が 10 億（a thousand million）で，英用法では a million million が 1 兆を意味していたためである（現在は米用法が一般的となっている）．

● 角度や年号などの 3～4 桁の数字は，百と十の位を区切って読むこともある．

360 → three sixty　　　　2016 → twenty sixteen

(2) 序数詞のルール

●数字で表す場合には，数の後ろに2文字（-th など）を添え，定冠詞 the をつける．

the first［1st］，the second［2nd］，the third［3rd］，the fourth［4th］，…，the twentieth［20th］，the twenty-first［21st］，…，the one hundredth［100th］，the one thousandth［1000th］

●通常，13～19 の語尾（-teen）に発音のアクセントがある．

●20 以上では，最後の桁だけを序数詞にする．

(3) 倍数詞のルール

●原則として，「～ times（as ... as）」を用いるが，2倍や2回を表すには twice や double，1回や1度を表すには once も使われる．

The cost rises twice as high as last year.「昨年の2倍に価格高騰した」

The magnitude reached double the predicted value.「予測値の倍の強度に達した」

●半分を表すには half を用いる．

Half（of）the test specimen was consumed one and a half years ago.「1年半前に試験片の半分を消耗させた」

6.2 小数と分数（Decimal fraction and Fraction）

●小数は decimal と呼び，小数点は decimal point と呼ぶ．数字中で小数点は単に point と読み，小数点以下の数字は日本語と同様に各位ごとに1桁の数字で読む．

3.14 → three point one four

0.09 → zero [nought] point zero nine または，o（オー） point o（オー） nine

●分数は fraction と呼び，基数詞表記の分子（numerator）と序数詞表記の分母（denominator）をハイフンでつなげて書き表す．日本語と異なり「分子→分母」の順で読む．分子が2以上の場合，分母は序数詞の複数形（-s）にする．

$\frac{1}{2}$ → a half　または　one-half　　$\frac{1}{3}$ → one-third　または　a third

$\frac{1}{4}$ → a quarter　または　one-forth　　　$\frac{2}{3}$ → two-thirds

●帯分数では分数の前に and を加えて読むが，技術英語において帯分数表記は一般的でない．ただし，複雑な分数や分母が大きい数字の場合，基数詞表記の分母に over をつけて読むことが多い，

$3\frac{1}{2}$ → three and a half　　　$2\frac{6}{7}$ → two and six-sevenths

$\frac{20}{7}$ → twenty over seven（ただし twenty-sevenths では 27th と誤解しやすい）

$\frac{1}{256}$ → one over two hundred（and）fifty-six

$\frac{640}{512}$ → six hundred forty over five hundred（and）twelve

6.3 数詞を含む表現（Expression of numerics）

① **時刻の読み方** ➡ 10：00 am（午前 10 時）→ ten　または　ten a m（エイ エム）
10：00 pm（午後 10 時）→ ten　または　ten p m（ピー エム）
09：55 → nine fifty-five　または　five（minutes）to ten
10：05 → ten five　または　five（minutes）past ten
10：30 → ten thirty　または　half past ten

② **日付の読み方** ➡ 2016 年 4 月 15 日 → April the fifteenth, twenty sixteen
H20 年 7 月 1 日 → July the first, the twenty year of Heisei

③ **金額の読み方** ➡ ¥2,000 → two thousand yen
€5.20 → five euro twenty（cent）

④ **温度の読み方** ➡ 4℃ → four degrees by centigrade［Celsius］「摂氏 4 度」
−5°F → minus five degrees Fahrenheit「華氏マイナス 5 度」
300 K → three hundred Kelvin「絶対温度 300 ケルビン」

⑤ **割合の表し方** ➡ 75％ → seventy-five per cent
0.25％ → zero point two five per cent

⑥ **不特定多数の表し方** ➡ hundreds of papers「何百もの論文」
millions of people「何百万人もの人々」

時間・距離・量などのまとまった単位を表すときは，名詞は複数形（-s）を用いるが，単数形の動詞で受ける．

Ten minutes is needed to melt it.「それを溶かすには 10 分必要だ」

Two meters is enough to mount the apparatus.「その装置の実装には 2m もあれば十分だ」

名詞を「数字と単位」で修飾する場合は，（必須ではないが）ハイフンで数字–単位間をつなぎ，単位は単数形（形容詞扱い）で表記する．

a 2-cyle engine

a 10,000-volt electrostatic charge

an 8-meter channel flow

6.4 数式の読み方（How to read equations）

等号を含む数式は一つの文章として読まれる．通常，等号を動詞として扱い，数式の左辺第一項（文頭の数字または文字）が主語となる．左辺（left-hand side）と右辺（right-hand side）を合わせて，両辺を both sides と呼ぶ．項を term と呼び，第一項は the first term と読むように序数詞が用いられる．

●加算（addition），減算（subtraction），乗算（multiplication），除算（division）の四則演算（four arithmetic operations）を例に，数式の読み方を以下に示す．

$x+y=z.$ → x plus y equals z.

文頭の x が主語であり，plus は前置詞，equals が三人称単数の -s を付与した動詞である．plus の代わりに and を用いた場合は，複数扱いでもよい．文の終わりとして，数式の末尾にはピリオドがつく．動詞の equal の代わりに is equal to を，または口語的には makes を使うこともでき，これは下記の減算などの場合も同等である．

$x-y=z.$ → x minus y equals z.

負の値を表す場合も minus を用いて，-10 は minus ten と読む．口語的には makes の他に，leaves（～を残す）も使われる．乗算と除算は，

$x\times y=z.$ → x times y equals z.

または　x multiplied by y equals z.

$xy=z.$ → xy equals z.

$x/y=z.$ → x over y equals z.

$x \div y = z.$ → x divided by y equals z.

と読まれる．以上の四則演算の読み方として，より口語的には，

$x + y = z.$ → If you add y to x, you get ［have］ z.

$x - y = z.$ → If you subtract ［take］ y from x, you get ［have］ z.

$x \times y = z.$ → If you multiply x by y, you get ［have］ z.

$x \div y = z.$ → If you divide x by y, you get ［have］ z.

と表現することもできる．you の他に we でもよいが，常に主語は必要である．日本では除算記号に「÷」がよく使われる一方，技術英語では一般的でなく分数と「/(slash)」記号が用いられる．

●不等式（inequality）も，等式（equality）と同様に一つの英文として読む．be 動詞による表現が一般的である．

$x < y.$ $[x > y.]$ → x is less ［greater］ than y.

$x \leq y.$ $[x \geq y.]$ → x is less ［greater］ than or equal to y.

$x < y \leq z.$ → x is less than y is less than or equal to z.

$x \neq y.$ → x is not equal to y.

$x \approx y.$ → x is approximately equal to y.

$x \equiv y.$ → x is equivalent to y. または　x is defined as y.

●式中に使われる括弧には種類があり，（丸括弧）は parenthesis，［角括弧］は bracket，{波括弧} は braces，〈山括弧〉は angle bracket と呼ばれる．通常，数式を読み上げる際には無視してよいが，

$(x + y)^2$ → x plus y squared

$x + y^2$ と区別したいなど，特に括弧を強調する場合には，括弧のはじめ（open）と括弧閉じ（close）も明確にして読むとよい．

$(x + y)^2$ → Open parenthesis x plus y close parenthesis squared

6.5 数列や極限の読み方（Sequence and limit）

●数列（sequence）は，

$a_0, a_1, a_2, \ldots a_i$ → a（sub）zero，a（sub）one，a（sub）two，... a（sub）i などの数の並びを指す．下添え字（subscript）を基数詞で読み，sub は省略されることもあり，a_0 は a naught と呼ぶこともある．上添え字（superscript）がついた a^i は a super i と読む．

●数列の総和や総乗は，

$\Sigma_{n=1}^{\infty} a_i$ → sum［summation］from n equals one to infinity of a sub i

$\Pi_{i=0}^{n} a_i$ → product from i equals zero to n of a sub i

と読む．「$n=1$ to ∞」などの範囲が指定されない場合は，単純に the sum［product］of ...と読めばよい．

●極限（limit）は，

$\lim_{n\to\infty} a_i$ → limit as n goes［tends］to infinity of a sub i

または　limit of a sub i as n goes［tends］to infinity

と読む．極限をとる部分（ここでは a_i）が長い場合は，前者の読みが好まれる．

6.6 関数の読み方（Function）

●関数（function）を読むとき，引数（argument）に of をつけて，

$f(x)$ → function f of x　または　f of x

と読む．例えば，$y=f(x)$ のとき，y が従属変数（dependent variable），x が独立変数（independent variable）と呼ばれる．これを微分する（differentiate）と得られる導関数（derivative）は，

$\frac{\mathrm{d}f(x)}{\mathrm{d}x}$ → derivative of f of x with respect to x　または　d（ディー） f（エフ） d（ディー） x（エックス）

と読む．「x について微分」を言い表すには，「with respect to x」を用いる．口語的には，簡略化されて分子→分母の順で文字の通りに読むことができる．

●2 階導関数などの高次（high-order）の微分や偏微分（partial differentiation）の場合は，たとえば，

$\frac{\mathrm{d}^2 f(x)}{\mathrm{d}x^2}$ → second-order derivative of f of x with respect x

$\frac{\partial f}{\partial x}$ → partial derivative of f with respect x

$\frac{\partial^2 f}{\partial x \partial y}$ →（mixed）second partial derivative of f with respect x and y

と読む．導関数の簡易表記として，「′（prime）」記号が用いられる．

$f'(x)$ → f prime of x　　$f''(x)$ → f double prime of x

$f'''(x)$ → f tripe prime of x　　$f^{(n)}(x)$ → f n prime of x

●積分する（integrate）場合も，積分変数をしているときは「with respect to」を利用し，

$\int f(x)\,\mathrm{d}x$ → integral（with respect to x）of f of x

$\int_a^b f\mathrm{d}x$ → integral（with respect to x）from a to b of f

$\int_V f\mathrm{d}x$ → integral over V of f

$\iint_V f\mathrm{d}x\mathrm{d}y$ → double integral over V of f of x and y

●べき（power）の例として，整数 n に対して，

x^n → x to the n-th（power）

5.02×10^3 → five point zero two times ten to the third（power）

と読み，指数（exponent）n の数は序数詞で扱われる．

2乗と3乗は例外として，

x^2 → x squared

$(x+y)^3$ → x plus y cubed

と読まれることもある．n が非整数や式の場合は，

x^{n+1} → x to the（power of）n plus one

$e^{1.5}$ → e to the（power of）one point five

$e^{-\frac{5}{3}}$ → e to the（power of）minus five-third

と読むことができ，基数詞の扱いになる．動詞 raise を用いた表現もある．

$e^{-\frac{5}{3}}$ → e raised to the minus five-third power

●対数（logarithm）には，ネイピア数（Napier's number）を底（base）とする自然対数（natural logarithm）と，10を底とする常用対数（common logarithm）がよく用いられる．

$\log_a x$ → log a x　または　log to the base a of x

$\ln x$ →（natural）log（of）x

$\log x$ →（common）log（of）x

●べき乗根（root）で，特に「√」根号記号（radical symbol）を用いた数式は，

$\sqrt[n]{x}$ → the n-th root of x

と読む．根号指数（radical index）n が 2 や 3 のとき，それぞれ平方根（square root）と立方根（cube root）と呼び，

$\sqrt{1-x^2}$　the square root of one minus x squared

$\sqrt[3]{xyz}$　the cube root of x y z

と読む．

6.7 集合と論理（Set and logical opration）

●集合（set）と元（element）の帰属関係（membership relation）の読み方を示す．

$a \in A$ → a is an element of A.　　$A \ni a$ → A contains a.

$A \subset B$[$A \supset B$] → A is a subset [superset] of B.

$A \cap B$[$A \cup B$] → A is the intersection [union] of B.

●命題（proposition）に関する論理演算式（logical operation）の読み方を示す．

$A \to B$ → A implies B.

$A \wedge B$[$A \vee B$] → conjunction [disjunction] of A and B

●実数の部分集合で，一つながりの範囲を表す区間（interval）について，閉区間と開区間などは以下のように読む．

$[a, b]$ → closed interval from a to b

(a, b) → open interval from a to b

※座標（a, b）の読み方は日本語同様「エービー」

$[a, b)$ → right-open interval from a to b

$(a, b]$ → left-open interval from a to b

6.8 図形，ベクトル（Diagram and vector）

幾何学（geometry）やベクトルの取扱い時に用いる用語について，以下に紹介する．単語 vector の発音は“ベクタァ”（ベクトルではなく）であり，対となる単語 scalar は“スケイラァ”（スカラーではなく）と読むことに注意する．

60° → sixty degree angle

∠ABC → angle A B C

△ABC → triangle A B C

$\overline{\mathrm{AB}}$ → line segment A B

$\overrightarrow{AB}$ → vector A B

AB⊥CD → line AB is perpendicular［orthogonal］to line CD

AB∥CD → line AB is parallel to line CD

△ABC≡△DEF → triangle A B C is congruent to triangle D E F

△ABC∽△DEF → triangle A B C is similar to triangle D E F

$\vec{a}$ → vector a

$|\vec{a}|$ → the magnitude of vector a

$|a|$ → the absolute value of a

$\vec{a} \cdot \vec{b}$ → inner product of vectors a and b　または　(vector) a dot b

$\vec{a} \times \vec{b}$ → cross product of vectors a and b　または　(vector) a times b

$\vec{a} \otimes \vec{b}$ → outer product of vectors a and b

7 技術系英文法の応用

Technical writing

7.1 技術英文で大事な要素

技術文章は，シェイクスピアによる文学的な文章とは，その作文の目的も特徴も異なる．文学では，文章で表現された世界観を読み手が想像して楽しみ，読み手による解釈・受取りはさまざまで自由である．ときに難解で，ときに平易で面白おかしく書かれていて，それが新たな余興や娯楽を生み出していく．必ずしも著者のイメージ通りに，読み手が受け取らずともよい．

一方で，技術文章は，その対極にある．たとえば，製品説明書やマニュアルでは，確実に品物や使用方法の詳細を，**過不足なく一切の誤解を与えない**ように伝えなければならない．読み手（使用者）によって説明書の解釈が異なるようでは，誤解を与えていることと同じである．説明書の誤解を原因にして，事故や損害が生じた場合には，その補償は書き手（製造者）が負わなければならない．企業間の契約についても同様のことがいえる．賠償問題になった際，契約内容の英文の解釈によって立場が変わることもある．つまり，技術文章は，法的責任が生じることや，大きな代償（損害賠償）に繋がることがあるため，**誠実に作文**していかなければならない．

学術論文や技術報告も，上述の技術文章の一種である．論文の主たる目的は，著者の発見や技術などを公知することである．その際，**論理的な文章構成**でなければ，論文とはいえない．偏見ばかりの文章や独りよがりの文章では，他者を納得させることはできない．誰もが理解し納得できる**客観的な文を一つ一つ丁寧に**書き，それを連続させて，文章，段落，章，そして論文へと積み（書き）上げていく．また，学術論文では新しい知識を，技術報告では新しい技術を公知して，他者にそれらを活用してもらうことが論文発表による目的でもある．そのためには，**透明性の高い文章**であることが大事である．たとえ有名な学術雑誌に掲載された論文でも，第三者によって再現性が認められない場合には，不正な研究として捏造（ねつぞう）・改竄（かいざん）の疑いがかかることもある．要らぬ嫌疑をかけられないためにも，透明性・客観性を意識して技術英文を執筆しなければならない．

以上のことは，技術英文に限らず日本語ベースでも，肝に銘じるべきことである．しかし，技術英文に不慣れなうちは，上述の誠実性・透明性・客観性を意識しながら作文していくことは難しい．そこで，以下では，3つのキーワードを添えて，より具体的な注意点を説明していく．

7.2 技術英文3つのキーワード“3C”

技術英文を書くときに意識すべき3つのキーワードは，各英単語の頭文字をとって「**3つのC**」と呼ばれている．

❶ **Correct（正確に書く）**➡ 文章全体への信頼を与えよう．

❷ **Clear　（明確に書く）**➡ 一通りの解釈で，誤解の余地をなくそう．

❸ **Concise（簡潔に書く）**➡ 読み手に容易に素早く読んでもらおう．

例文を用いて，各キーワードの意味とテクニックの一例を紹介しよう．

正確な（correct）文章は技術英文で必要不可欠で，逆に不正確な文章があってはならない．次の英文は「NC工作機械は高精度である」を訳したものであるが，文①は間違いである（文の内容を無視すれば，文法自体は間違っていない）．

① The NC machine tools are high accuracy.

② Accuracy of the NC machine tools is high.

③ The NC machine tools operate with high accuracy.

文①は日本語に直訳しても違和感はないかもしれないが，これは日本人にしか通じにくい（日本語に引きずられた）英文である．文の構成を見ると，machine tools と are と accuracy が SVC 文型でつながっているが，“machine tools are accuracy”としてみると違和感を感じとれよう．これは，machine tool（工作機械）が accuracy（精度）そのもの（主格補語）にはならないためである．high accuracy「高精度」となっても，形容詞 high が名詞 accuracy を修飾したのみで，「高精度“な”」とはならないことに注意すべきである．

よって，文の意味がより正確に伝わるよう，文②や文③のように書くべきである．さらに，主部の長い文②より，SVO 型の文③のほうが直接的で伝わりやすい文章である．正確な英文を書くには，文法と単語の誤りはもちろん，動詞の選択，数や数値の表記法などにも気をつけること．

明確な（clear）文章は，誰が読んでも同じように解釈され，正確な（correct）文章作りにもつながる．「十分に冷却せよ」という文章も，具体的にものごとを書くことで，明確に書き手の意図を伝えられる．

Cool it sufficiently.

上の文では，どれほど冷やせばよいかがわからない．たとえば，次の2文のように，具体的な数字や，背景にある目的を述べれば，読み手も理解しやすい．

Cool it down to a temperature of 0℃.「0度まで冷却せよ」

Make it cold enough to touch.「手で触れられるまで十分に冷却せよ」

簡潔な（concise）文章，つまり短くて平易な文にすることで，正確さと明確さも向上させやすくなる．ネイティブが書くような凝った表現を真似ようとしたり，たくさんのメッセージを1文に詰め込もうとしたりと，日本人の英文は冗長的なものになりがちである．しかし，そのような文章は読みにくく，誤解も生じやすい．

技術英文における主目的は，**ただ事実・メッセージを読み手に淡々と伝えること**であり，**凝った言い回しは不要**である．読みやすく，わかりやすい文章が好まれる．次の2文は，いずれも「実験で速度の測定を行った」の英訳である．

① We make a measurement of the velocity in experiment.

② We measure the velocity (in experiment).

文①は与えられた日本語を丁寧に（単語毎に）英訳しており，結果的に冗長な文章になっている．make a measurement of「～の測定を行う」の表現は間違いでないが，「～を測定する」を言い表す専用の動詞 measure があり，これを活用すれば大幅な語数削減につながる．さらに，「測定」自体が「実験」の一種とも認識できるのだから，わざわざ in experiment を付す必要性もない．日本語の文章に引きずられ過ぎず，各文の内容を見極め，文②のように本質のみを英語で書き表していけばよい．

以上，3つのCを心掛けていけばよいが，さらに具体的なテクニックを以下に述べていく．前述の「**日本語の文章に引きずられ過ぎず**」，「**専用の動詞を活用していくこと**」などが第一歩である．

7.3 日本語的発想，動詞を活かすSVO文型の使用

日本人が不慣れなうちに英文を書く際，元の日本語の文章に引きずられてしまい，不自然な英文や冗長的な文章になることが多い．これは開発途上分野である翻訳ソフトウェアを使ってみても，同様である．たとえば，「本工程を経ることで，多くの費用と時間が発生する」を某ソフトウェアの翻訳に頼ると，

① By going through this process, a lot of cost and time occur.

② A lot of cost and time generated by going through this process.

と翻訳結果が返ってきた．文②は，そもそも文の体を成していないが，文①も冗長な文章となっている．ここで，「(お金や時間) を消耗する」を言い表す専用の動詞 consume を用いて，SVO 文型に書き換えると，

This process consumes cost and time.

のように，簡潔で明快な表現となる．

日本語では「○○をする」の言い回しが多いため，これを英語に直訳するとき「○○」を名詞にして，do，have，give，make などの広範に使える動詞を利用してしまう．これが日本語に引きずられる症状の一例である．

前述の consume など，より狭い（特化した）意味を持つ動詞を活かし，「誰が・何を・どうする」の SVO 型で直接的に伝えるとよい．SVO 型は五文型の中でも，技術英文に頻出する最も直接的で強い印象を与える型である．以下に，ほかの文例を示す．

「以下の図は，装置の概略図を示す」➡

① The following figure shows a schematic illustration of the device.

② The following schematically illustrates the device.

文②では，動詞 illustrate「～を説明（解説）する」を活かし，SVO 型で少ない語数で表現している．文①の figure と illustration は両方とも「図」を指しており，冗長である．次の，文例は多用しがちな be 動詞を書き換えたものである．

「銅は熱の伝導体で，電気伝導体である」➡

① Cupper is a conductor of heat and a conductor of electricity.

② Cupper conducts heat and electricity.

7.4 無生物主語の活用

英語表現では，必ず文章中に主語が含まれている．しかし，学術論文や技術報告などでは，動作主（人物）を明らかにする必要のないことが多い．むしろ事象や物が主語・主役となって文章が構成されているほうが自然であるかもしれない．つまり，技術英文では無生物主語を積極的に活用するように心掛けると，余計な情報を含まない簡潔で明確な文章となる．以下に，例を示す．

「その試験片のヤング率を引張り試験で測定した．」➡

Ⅵ　文　　法

① We measured the Young's modulus of the specimen with a tensile test

② The Young's modulus of the specimen was measured with a tensile test.

文②では，「ヤング率を引張り試験で測定した」ことに重点が置かれているが，文①では，誰が測定したかも伝えているため，強調部分がややぼやけている．もし，たとえば「(先行研究がなく，初めて) 私達が行った」ことを強調したいのであれば，文①が効果的であろう．

「流量調整には弁が必要である」➡

① We need a valve for the flow control.

② The flow control requires a valve.

文①の we が「(一般的な) 人々は」のニュアンスであるならば，we の必要性は弱く，文②のように動詞 require を活用した SVO 型で表す方法が効果的である．

問題解答

Ⅰ　機械工学の基礎

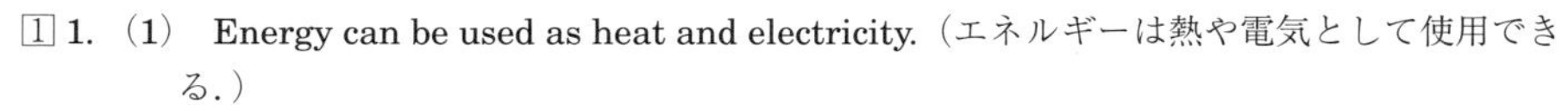

[1] 1. (1)　Energy can be used as heat and electricity.（エネルギーは熱や電気として使用できる.）

(2)　Chemical energy is transformed into electrical energy.（化学的エネルギーは電気エネルギーに変換される.）

(3)　Electrical energy results from the motion of electrons.（電気エネルギーは電子の運動で生じる.）

2. (1)　○　(2)　×　(3)　×　(4)　×

[2] 1. (1)　Heavy bodies fall faster than light ones.（重い物体は軽いものよりも早く落ちる.）

(2)　He actually carried out experiments to test the theory.（彼は実際にその理論をテストするため実験を行った.）

(3)　The experimental method has become very important.（実験的な方法は重要になってきた.）

2. (1)　×　(2)　×　(3)　○　(4)　○

[3] 1. (1)　Rubbing the palms of your hands together generates warmth.（あなたの手のひらをこすりあわせることは暖かさを生み出す.）

(2)　Friction prevents the wheels of automobiles from slipping.（摩擦は自動車の車輪を滑るのを防ぐ.）

(3)　Lubricant is usually applied to reduce friction.（潤滑剤は通常，摩擦を減らすために使用される.）

2. (1)　○　(2)　○　(3)　○　(4)　×

[4] 1. (1)　A bent lever is useful in a machine.（曲がったレバーは機械で役に立つ.）

(2)　The point holding a lever is called a fulcrum.（レバーを保つ点は支点と呼ばれる.）

(3)　We can calculate the power exerted on the nail.（私たちはくぎに働く力を計算することができる.）

2. (1)　×　(2)　×　(3)　○　(4)　○

[5] 1. (1)　You inject high pressure air into a flat tire.（あなたはパンクしたタイヤに高圧の空気を注入する.）

(2)　The driver must have stepped on his brake pedal.（運転手は彼のブレーキペダルを踏んだに違いない.）

(3)　The movement of the car body is stopped by an immense pressure.（車体の動きは巨大な圧力によって止められる.）

2. (1)　○　(2)　×　(3)　○　(4)　×

問題解答

6 1. (1) Electricity flows along a copper wire like water.（電気は水のように銅のワイヤに沿って流れる.）

(2) The conductor can conduct electricity from one place to another.（導体はある場所から別の場所へ電気を伝導することができる.）

(3) The insulator has very poor conductivity.（絶縁体は非常に低い伝導率をもつ.）

2. (1) ○ (2) × (3) × (4) ×

7 1. (1) Electricity flows from one place to another.（電気はある場所から別の場所へ流れる.）

(2) An electric circuit should be a closed circuit.（電気回路は閉回路でなければならない.）

(3) A closed circuit is necessary to obtain work.（仕事を得るためには閉回路が必要である.）

2. (1) ○ (2) ○ (3) × (4) ○

8 1. (1) An internal combustion engine is most commonly used in automobiles.（内燃機関が自動車で最も一般的に使われる.）

(2) An internal combustion engine consists of cylinders and pistons.（内燃機関はシリンダとピストンからなる.）

(3) The exhaust gas contains contaminating substances.（排気ガスは汚染されている物質を含む.）

2. (1) ○ (2) × (3) × (4) ○

9 1. (1) The metric system is a system used in almost every country.（メートル単位系はほとんどあらゆる国で使われる単位系である.）

(2) The common prefixes used for calculations are milli and kilo.（計算のために使われる普通の接頭辞はミリとキロです.）

(3) 2000 meters is usually called 2 kilometers.（2000 メートルは普通 2 キロメートルと呼ばれる.）

2. (1) ○ (2) × (3) ○ (4) ○

10 1. (1) There are several reasons for the measurement error.（測定誤差にはいくつかの理由がある.）

(2) It is necessary to indicate the degree of uncertainty.（不確実性の程度を示すことが必要である.）

(3) The "±" sign is used to express the uncertainty in a measurement.（「±」のサインは測定において不確実性を表すのに用いられる.）

2. (1) ○ (2) ○ (3) × (4) ○

11 1. (1) There is no relation between the number of significant figures and the decimal point.（有効数字の数と小数点の間には関係がない.）

(2) The power system is used to clarify the number of significant figures.（乗べき記法が有効数字の数をはっきりさせるのに用いられる.）

(3) The number of significant figures of 5020 can' t be easily judged.（5020 の有効数

字の数は簡単に判断できない．）

2. (1) × (2) ○ (3) ○ (4) ×

12 1. (1) A functional relation is found in various properties of substances.（関数関係は物質のいろいろな特性で見つかる．）

(2) The relation between gas pressure and its volume is an inverse proportion.（気体の圧力とその体積の関係は反比例である．）

(3) An inverse proportion is expressed mathematically as $y=a/x$.（反比例は $y=a/x$ として数学的に表される．）

2. (1) × (2) × (3) ○ (4) ×

Ⅱ 機械工学の周辺

1 1. (1) The worker must have information on its precise shape and dimensions.（作業者はその正確な形と寸法に関する情報をもってなければならない．）

(2) A pictorial drawing is not suitable to obtain the exact shape and dimensions.（見取図は正確な形と寸法を得るために適当でありません．）

(3) We require at least two projections on the object to be made.（私たちはつくるべき物体に関する少なくとも二つ以上の図面を必要とする．）

(4) The working drawings must be clear and understandable.（製作図は明白で理解できなければならない．）

2. (1) × (2) ○ (3) × (4) ○

2 1. (1) A visible line is a continuous straight thick line.（外形線は連続したまっすぐな太い線である．）

(2) A dimension line has arrow marks at both ends.（寸法線は両端に矢印をもつ．）

(3) A hidden line is a thick line composed of dashes and spaces.（かくれ線はダッシュとすき間からなる太い線である．）

(4) An extension line is a continuous straight thin line.（寸法補助線は連続したまっすぐな細い線である．）

2. (1) ○ (2) × (3) ○ (4) ×

3 1. (1) A micrometer is a precision instrument to measure small distances.（マイクロメータは小さい距離を測定するための精密器機である．）

(2) A micrometer must be handled as carefully as possible.（マイクロメータはできるだけ慎重に取り扱われなければならない．）

(3) A micrometer may be damaged by dropping it on the floor.（マイクロメータは床にそれを落とすことによって被害を受けるかもしれない．）

(4) A micrometer must be stored in the case to prevent contamination.（マイクロメータは不純物の混入を防ぐためケースの中に保存されなければならない．）

2. (1) ○ (2) ○ (3) ○ (4) ×

4 1. (1) A gauge is a device to determine the size of an object.（ゲージは物の大きさを測定

する装置である．）

（2） A gauge block is one of the typical end standard.（ゲージブロックは典型的な端度器の一つである．）

（3） Jo-gauge is used for calibrating a graduated gauge.（ジョー・ゲージは目盛のついたゲージを校正するために使われる．）

（4） The final standard of length is the wave length of red light.（長さの最終的な標準は赤い光の波長である．）

2. （1） ○ （2） ○ （3） × （4） ×

5 1. （1） A limit gauge is a gauge to test the size of the workpiece.（限界ゲージは製品の寸法を検査するためのゲージである．）

（2） The size of the workpiece should be between the upper limit and the lower limit.（製品の大きさは上限と下限の間になければならない．）

（3） A limit gauge system is essential to mass manufacturing.（限界ゲージシステムは大量生産にとって必須である．）

（4） A limit gauge system has a close relation to interchangeability.（限界ゲージシステムは互換性と緊密な関係がある．）

2. （1） × （2） ○ （3） ○ （4） ×

6 1. （1） It is important to know the limit of endurance of materials to exterior forces.（外力に対する材料の耐力の限界を知っていることは重要である．）

（2） A stress is obtained by dividing the load by the original area of the cross section.（応力は最初の断面積で荷重を割ることによって得られる．）

（3） An elastic range is the range where the strain is proportional to the stress.（弾性範囲はひずみが応力に比例している範囲である．）

（4） A yield point is the point where strain increases without an increment of stress.（降伏点はひずみが応力の増加なしで増加する点です．）

2. （1） ○ （2） × （3） ○ （4） ×

7 1. （1） Tensile strength refers to the maximum stress a metal can bear before breaking.（引張強さは金属が壊れる前に耐えることができる最大の応力を意味する．）

（2） Hardness refers to the extent that the metal can resist to plastic deformation.（硬さは金属が塑性変形に抵抗することができる程度を意味する．）

（3） Ductility refers to the extent that the metal can be drawn out without breaking.（延性は金属が壊れることなく引き延ばすことができる程度を意味する．）

（4） Malleability refers to the extent that the metal can be bent without cracking.（可鍛性は金属がひび割れることなく曲げられることができる程度を意味する．）

2. （1） × （2） ○ （3） × （4） ○

8 1. （1） An alloy is a metal mixture containing two or more metals.（合金は二つまたはそれ以上の金属を含む金属の混合物である．）

（2） An alloy is more common than a pure metal.（合金は純粋な金属より一般的である．）

(3) An alloy is harder than its constituent metals.（合金はその構成金属より硬い.）

(4) An alloy is made of melting and mixing two or more metals.（合金は二つまたはそれ以上の金属を溶かして混ぜることでできている.）

2. (1) × (2) ○ (3) ○ (4) ○

9 1. (1) Shape-memory alloys have a very interesting and useful property.（形状記憶合金は非常に面白くて役に立つ特性をもつ.）

(2) Shape-memory alloys give us very big convenience.（形状記憶合金は我々に非常に大きい便利さを与える.）

(3) This effect results from the transformation of the martensite phase.（この効果はマルテンサイト相の変態によって生じる.）

(4) The bent wire would straighten if it is heated.（その曲がったワイヤーはそれが熱されるならばまっすぐになるだろう.）

2. (1) ○ (2) × (3) × (4) ×

10 1. (1) Composite materials are usually composed of two or more materials.（複合材料は普通二つまたはそれ以上の材料から構成される.）

(2) Composite materials are made by mixing and bonding two or more materials.（複合材料は二つまたはそれ以上の材料を混ぜて結合することによって作られる.）

(3) Composite materials can increase their resistance to fatigue and corrosion.（複合材料は疲労と腐食に対するそれらの抵抗力を増加することができる.）

(4) A new advanced composite material consists of a fiber and a mother material.（新しい先進複合材料は繊維と母材からなる.）

2. (1) × (2) ○ (3) × (4) ○

11 1. (1) Carbon fiber is used as a strengthening material.（炭素繊維は強化材料として使われる.）

(2) Carbon fiber is a material consisting of extremely thin fibers.（炭素繊維は極端に細い繊維からなる材料である.）

(3) FRP is the most famous composite material.（FRP は最も有名な複合材料である.）

(4) FRP can be found everywhere from tennis rackets to the aerospace industry.（FRP はテニスラケットから航空宇宙産業まで至る所で見つかる.）

2. (1) ○ (2) × (3) ○ (4) ○

12 1. (1) Crude petroleum is a mixture of various grades of hydrocarbons.（原油は炭化水素のいろいろな等級の混合物である.）

(2) It is separated by distillation into various fractions.（それは蒸留によっていろいろな成分にわけられる.）

(3) Hydrocarbons from butane to C_{12} compounds appear in fuel.（ブタンから C12 化合物までの炭化水素は燃料の中に現れる.）

(4) Residues can be used to yield familiar products as asphalt and coke.（残留物はアスファルトやコークスのようなよく知られた製品をつくるのに使うことができる.）

2. (1) × (2) ○ (3) ○ (4) ○

Ⅲ 機械工作

1 1. (1) A machine has rigid bodies forced to make specific motions.（機械は特定の動作をすることを強いられるしっかりとした本体をもつ.）
(2) A machine is capable of performing useful work.（機械は役に立つ仕事を実行することができる.）
(3) A machine must receive energy and deliver work in a useful form.（機械はエネルギーを受けとり，そして役に立つ形で仕事を伝達しなければならない.）
(4) A machine is defined as a device that converts energy or information.（機械はエネルギーまたは情報を変換する装置として定義される.）
2. (1) × (2) ○ (3) × (4) ○

2 1. (1) A lathe is a machine tool used to shape a piece of metal.（旋盤は金属片をつくるために使用される工作機械である.）
(2) A lathe is the most important machine tool.（旋盤は最も重要な工作機械である.）
(3) A workpiece is a material shaved by the lathe.（工作物は旋盤で削られる材料である.）
(4) A workpiece must be securely held by a chuck before cutting.（工作物は切削の前にチャックでしっかりと保持されなければならない.）
2. (1) × (2) × (3) ○ (4) ○

3 1. (1) A milling machine usually uses a multiple point cutting tool.（フライス盤は普通多点切削工具を使う.）
(2) Milling machines are classified into three different types.（フライス盤は三つの異なるタイプに分類される.）
(3) Milling machines provide a wide-range of operations for cutting.（フライス盤は切削のための広範囲の操作を提供する.）
(4) A complex curved surface can be prepared by a milling machine.（複雑な曲面はフライス盤でつくることができる.）
2. (1) × (2) × (3) × (4) ○

4 1. (1) A drilling machine is called a drill press.（穴をあける機械はボール盤と呼ばれる.）
(2) A drill press is a machine to make holes in metal.（ボール盤は金属に穴を作る機械である.）
(3) A twist drill is a drill with two twisted flutes to remove shavings.（ツイストドリルは削りくずを除去するために二つのねじれた溝付きのドリルである.）
(4) A drill jig is a special tool for holding the workpiece.（ドリルジグは工作物を保つための特別な道具である.）
2. (1) × (2) ○ (3) × (4) ○

5 1. (1) A grinding machine is called a grinder for short.（研削盤は略してグラインダーと

呼ばれる.）

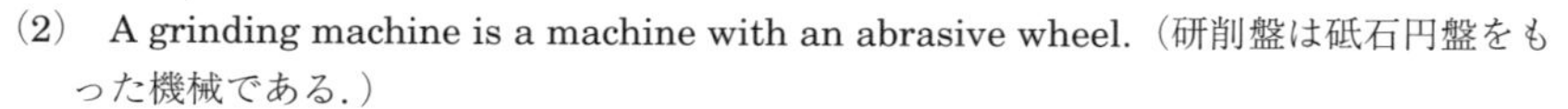

（2） A grinding machine is a machine with an abrasive wheel.（研削盤は砥石円盤をもった機械である.）

（3） There are two types of grinding machines.（研削盤には二種類ある.）

（4） A magnetic chuck is the tool used to hold the workpiece.（マグネットチャックは工作物を保持するのに用いられる道具である.）

2. （1） × （2） ○ （3） × （4） ○

6 1. （1） There is a relation among the shear angle, the rake angle, and the cutting ratio.（せん断角，すくい角，そして切断速度の間に関係がある.）

（2） The tool moves along the surface of the workpiece to remove the surface.（工具は表面を取り除くために工作物の表面に沿って動く.）

（3） The material in front of the tool is continuously sheared to make chips.（工具の前の材料はチップを作るために連続的にせん断される.）

（4） The shear angle can control the thickness of the chip.（せん断角はチップの厚さを制御することができる.）

2. （1） × （2） × （3） × （4） ○

7 1. （1） A lubricant is used to prevent direct contact of two parts in relative motion.（潤滑剤は相対運動をしている二つの部品の直接接触を妨げるのに用いられる.）

（2） A lubricant can reduce friction and wear of the parts.（潤滑剤は部品の摩擦と摩耗を減らすことができる.）

（3） A lubricant can prevent the parts from rusting.（潤滑剤は部品がさびるのを防ぐことができる.）

（4） A lubricant works by creating a thick oil layer between two parts.（潤滑剤は二つの部品の間で厚い油層をつくることによって働く.）

2. （1） × （2） × （3） ○ （4） ○

8 1. （1） Welding is often done by melting metal pieces to join them together.（溶接はそれらをくっつけるために金属片を溶かすことによってしばしばなされる.）

（2） The pieces of metal that will be welded are called the base metal.（溶接される金属片は母材と呼ばれている.）

（3） Forge welding is another way to join metal pieces.（鍛接は金属片をくっつけるためのもう一つの方法である.）

（4） Forge welding does not fuse the metal pieces that will be joined.（鍛接は接合される金属片を溶かさない.）

2. （1） ○ （2） × （3） × （4） ○

9 1. （1） Forging refers to the plastic deformation of metals.（鍛造は金属の塑性変形を意味する.）

（2） Forging is a process that shapes metal by heating and hammering the metal.（鍛造は金属を熱して打つことによって金属を成形するプロセスである.）

(3) A forged metal has greater tensile strength and toughness.（鍛造された金属はより大きな引張強さとじん性をもつ.）

(4) Tools such as spanners and nippers are usually produced by forging.（スパナやニッパのような工具は普通鍛造によって生産される.）

2. (1) × (2) ○ (3) ○ (4) ×

10 1. (1) Casting is a process that creats something by pouring molten metal into a mold.（鋳造は溶解した金属を型へ注ぐことによって何かをつくるプロセスである.）

(2) The mold for casting is usually made from sand or metal.（鋳造のための型は普通砂か金属で製造される.）

(3) Casting offers several advantages over other metal forming process.（鋳造はほかの金属加工プロセスに勝るいくつかの利点を提供する.）

(4) Castings are objects which are made by casting.（鋳物は鋳造によってつくられた物体である.）

2. (1) × (2) ○ (3) × (4) ×

11 1. (1) The bloom is a mass of iron or steel to be rolled.（ブルームは圧延される鉄または鋼の塊である.）

(2) Rolling is a process where the bloom is passed through rolls.（圧延はブルームがロールを通されるプロセスだ.）

(3) Rolling is classified into two types by the rolling temperature.（圧延は圧延される温度によって二つのタイプに分類される.）

(4) Cold-rolled steel is made from hot-rolled steel.（冷間圧延鋼は熱間圧延鋼からつくられる.）

2. (1) ○ (2) × (3) × (4) ○

12 1. (1) Heat treatment involves heating and cooling of metal.（熱処理は金属を加熱したり冷却したりすることを意味する.）

(2) Heat treatment can alter the properties of the metal.（熱処理は金属の性質を変えることができる.）

(3) Carbon steel is particularly suited for heat treatment.（炭素鋼は特に熱処理に適している.）

(4) Tempering is similar to annealing in regard to removing brittleness.（焼き戻しはもろさを取り除くという点で焼きなましに似ている.）

2. (1) ○ (2) × (3) × (4) ○

Ⅳ 機械工学の現在

1 1. (1) CAD is an acronym for "computer-aided design".（CADは「コンピュータ援用設計」の頭字語である.）

(2) CAD refers to the use of computers for designing or planning systems.（CADはシステムを設計または計画するためにコンピュータを使用することを意味する.）

(3) The productivity has been remarkably improved by using CAD.（生産性はCADを使うことによって著しく改善された.）

(4) Drawing boards and instruments may be replaced by a personal computer.（製図板と器具はパソコンに置き替えられるかもしれない.）

2.（1） × （2） ○ （3） × （4） ○

2 1.（1） The motion in a machine was transformed through a physical mechanism.（機械の運動は物理的なメカニズムによって変えられた.）

(2) Mechatronics is a combined technology of mechanics and electronics.（メカトロニクスは機械工学と電子工学が組み合わされた技術である.）

(3) Robots are one of the typical examples for mechatronics.（ロボットはメカトロニクスの典型的例の一つである.）

(4) Electronic devices have played very important roles in mechatronics.（電子デバイスはメカトロニクスにおいて非常に重要な役割を演じてきた.）

2.（1） × （2） × （3） × （4） ×

3 1.（1） The devices that can be used instead of human senses are called sensors.（人間の感覚の代わりに使われることができる装置はセンサと呼ばれる.）

(2) A sensor receives a signal and responds it in a distinctive manner.（センサは信号を受け取り，そして特有の方法でそれに応答する.）

(3) A thermocouple is a sensor which converts temperature to voltage.（熱電対は温度を電圧に変えるセンサである.）

(4) A limit switch is a sensor used in electrical appliances.（リミットスイッチは電気器具で使われるセンサである.）

2.（1） ○ （2） ○ （3） ○ （4） ○

4 1.（1） A closed-loop control system has a sensor and a feedback route.（閉回路制御システムはセンサとフィードバック回路をもっている.）

(2) The route from the sensor point to the controller is called a feedback route.（センサ点からコントローラまでのルートはフィードバック回路と呼ばれる.）

(3) A feedback route can only transmit information.（フィードバック回路は情報を送ることだけができる.）

(4) A closed-loop control system has been applied in a refrigerator.（閉回路制御システムは冷蔵庫で適用されている.）

2.（1） × （2） ○ （3） ○ （4） ×

5 1.（1） NC in "NC machine tool" is the acronym for numerical control.（「NC工作機械」のNCは数値制御の頭字語である.）

(2) NC machine tool has not only the machine tool but also a computer.（NC工作機械は工作機械だけでなくコンピュータももっている.）

(3) In an old NC machine tool, the numerical values were stored in a magnetic tape.（古いNC工作機械では，数値は磁気テープに保存されていた.）

(4) Nowadays, almost all NC machine tools are computer numerical controlled.（今日，ほとんどすべての NC 工作機械はコントロール数値制御である.）

2. (1) ○ (2) ○ (3) × (4) ○

6 1. (1) A foreman had arranged for machines to be used and materials to be fed.（職長は使われる機械と供給される材料の手配をしていた.）

(2) CAM is a computerized system which can work in place of a foreman.（CAM は職長の代わりに働くことができるコンピュータ化されたシステムである.）

(3) The benefits brought about by CAM include increased productivity.（CAM によってもたらされる利益は増加した生産性を含む.）

(4) A foreman's one touch displays the most suitable rearrangement of personnel.（職長によるワンタッチが人員の最も適当な再編成を示す.）

2. (1) ○ (2) × (3) ○ (4) ○

7 1. (1) CIM is a system to optimize manufacturing activity.（CIM は製造活動を最適化するシステムである.）

(2) All informations are connected in a computerized network.（すべての情報はコンピュータ化されたネットワークでつながれる.）

(3) Functional areas are organically linked with each other.（機能的分野は互いに有機的に結合している.）

(4) A robot plays an important role in the CIM factory.（ロボットは CIM 工場で重要な役割を演じる.）

2. (1) ○ (2) × (3) ○ (4) ○

8 1. (1) The traditional factory equips with a rigid assembly line.（伝統的な工場は組立てラインを装備する.）

(2) FMS is a manufacturing system in which there is some amount of flexibility.（FMS はいくらか柔軟性がある製造システムである.）

(3) The line equips with these computerized devices.（ラインはこれらのコンピュータ化された装置を装備する.）

(4) This system was first introduced in machine shops.（このシステムは機械工場に最初に導入された.）

2. (1) ○ (2) ○ (3) × (4) ×

9 1. (1) The fuel cell generates electricity by a chemical reaction.（燃料電池は化学反応によって電気を発生させる.）

(2) Electricity is produced by the reaction between hydrogen and oxygen.（電気は水素と酸素の間で反応によって製造される.）

(3) Many types of fuel cells have been developed.（多くの種類の燃料電池が開発されてきた.）

(4) AFC was used for the famous spacecraft named *Apollo*.（AFC はアポロと呼ばれる有名な宇宙船のために使われた.）

2. (1) × (2) ○ (3) ○ (4) ×

10 1. (1) A solar cell generates electricity by the photovoltaic effect.（太陽電池は光電効果によって電気を発生させる.）

(2) A solar cell is made of semiconductors such as silicon.（太陽電池はシリコンのような半導体でできている.）

(3) A solar cell is very reliable because there are no moving parts.（太陽電池は可動部分がないので非常に信頼できる.）

(4) A solar cell cannot be used without sufficient light.（太陽電池は十分な光なしでは使うことができない.）

2. (1) × (2) ○ (3) ○ (4) ×

11 1. (1) The hybrid car has both a combustion engine and an electric motor.（ハイブリッド車は内燃機関と電気モーターの両方をもっている.）

(2) The hybrid car can drive more miles than other cars.（ハイブリッド車はほかの車より多くのマイルを運転することができる.）

(3) Regenerated brakes are brakes to generate electricity.（回生ブレーキは電気を発生させるブレーキである.）

(4) The hybrid car is more expensive than comparable cars without a motor.（ハイブリッド車はモータをもたない同等の車より高価だ.）

2. (1) × (2) ○ (3) ○ (4) ○

12 1. (1) The electric resistance of the superconductor becomes zero at low temperatures.（超伝導体の電気抵抗は低温でゼロになる.）

(2) This phenomenon has been known for more than a hundred years.（この現象は100年以上知られていた.）

(3) A linear motor car will be able to run between Tokyo and Nagoya in 40 minutes.（リニアモータカーは40分で東京と名古屋の間を走ることができるでしょう.）

(4) MRI is a superconducting device to diagnose conditions of the human body.（MRIは人体の状況を診断するための超伝導装置だ.）

2. (1) ○ (2) × (3) × (4) ○

V 管理技術

1 1. (1) An automobile consists of thousands of parts.（自動車は何千ものパーツからなる.）

(2) It is extremely important to make each part exactly interchangeable.（各部品を正確に互換性のあるよう製造することは極めて重要である.）

(3) They will be accurately assembled according to the blue print.（それらは青写真に従って正確に組み立てられるでしょう.）

(4) A worker must file individual parts to assemble them properly.（作業者はそれらを適切に組み立てるために個々の部品にやすりをかけなければならない.）

2. (1) ○ (2) × (3) × (4) ○

2 1. (1) Sampling inspection is the most powerful and effective method.（サンプリング検査は最も強力で効果的な方法である.）

(2) The inspection gives each product a little damage.（検査は各々の製品に少しのダメージを与える.）

(3) The most important factor is the method of selecting samples from the lot.（最も重要な要因はロットからサンプルを選ぶ方法である.）

(4) The method of selection is called random sampling.（選択の方法は無作為抽出と呼ばれる.）

2. (1) × (2) ○ (3) ○ (4) ×

3 1. (1) Inspection is to check the properties or dimensions of products.（検査は製品の特性または寸法をチェックすることである.）

(2) Measuring instruments have to be inspected regularly.（計測器は定期的に調べられなければならない.）

(3) Inspection is usually handled by the inspection department.（検査は普通検査部門によって取り扱われる.）

(4) The concept of inspection is a wider concept than the concept of testing.（検査の概念は試験の概念より広い概念である.）

2. (1) × (2) ○ (3) × (4) ×

4 1. (1) A control chart is used to evaluate the manufacturing process.（管理図は製造プロセスを評価するために用いられる.）

(2) A control chart usually consists of a central line and a pair of parallel lines.（管理図は普通，中央線と一対の平行線からなる.）

(3) The dimensions of the manufactured products are plotted on the chart.（製造された製品の寸法はチャートにプロットされる.）

(4) The theory of control charts is based on statistics.（管理図の理論は統計に基づく.）

2. (1) × (2) × (3) × (4) ○

5 1. (1) OR originated from the operational activity of the British army.（OR は英国軍の作戦行動から始まった.）

(2) OR applies scientific methods and techniques to determine the best solution.（OR は最高の解決策を決定するために科学的手法と技術を適用する.）

(3) OR has taught us that it is very important to make a suitable model.（OR は適切なモデルを作るすることが非常に重要であると我々に教えた.）

(4) In a sense, OR is similar to computer simulation.（ある意味では，OR はコンピュータシミュレーションと類似している.）

2. (1) ○ (2) × (3) × (4) ○

6 1. (1) System engineering is a branch of engineering.（システム工学は工学の一分野である.）

(2) System engineering involves the two operations of modeling and optimization.（シ

ステム工学はモデリングと最適化の二つの操作を含む.)

(3) They should be combined with the aid of a computer or mathematics.(それらはコンピュータまたは数学を用いて組み合わされるべきである.)

(4) A vast number of factors is integrated into a complicated system.(非常に多くの要因は複雑なシステムへ統合される.)

2. (1) × (2) ○ (3) ○ (4) ○

7 1. (1) The technology was always thought to have positive effects on society.(テクノロジーは常に社会に肯定的な影響を及ぼすと考えられた.)

(2) A negative evaluation is called technology assessment or TA for short.(否定的な評価はテクノロジーアセスメントまたは略して TA と呼ばれる.)

(3) In Japan, the first TA was for agricultural chemicals on golf links.(日本で，最初の TA はゴルフ・コースで農薬のためだった.)

(4) TA is a process to evaluate influence before introduction of a new technique.(TA は新しい技術の導入の前に影響を評価するプロセスである.)

2. (1) × (2) × (3) × (4) ○

8 1. (1) PERT is an acronym for program evaluation and review technique.(PERT は計画評価と確認の技術の頭字語だ.)

(2) PERT was developed for the Polaris missile program in the United States.(PERT はアメリカ合衆国でポラリスミサイル計画のために開発された.)

(3) PERT is one of the techniques in process planning and management.(PERT はプロセス計画と管理における技術の一つである.)

(4) PERT is applied to shorten the schedule for software development.(PERT はソフトウェア開発のスケジュールを短くするために適用される.)

2. (1) ○ (2) × (3) ○ (4) ×

9 1. (1) Nondestructive testing is to detect faults of products without damaging down.(非破壊試験は損害を与えることなく製品の欠陥を見つけることである.)

(2) Nondestructive testing is sometimes called nondestructive inspection.(非破壊試験は時々，破壊しない点検と呼ばれる.)

(3) Nondestructive testing is considered to be a very useful testing method.(非破壊試験は非常に役に立つ試験方法であると考えられる.)

(4) Nondestructive testing is a method to find flaws inside materials.(非破壊試験は材料の中の欠陥を発見する方法である.)

2. (1) × (2) ○ (3) ○ (4) ×

章末問題解答

Ⅰ 機械工学の基礎

1. (1) into(モータは電気エネルギーを機械的エネルギーに変える.)
   (2) than(コインは羽より速く落ちるかもしれない.)
   (3) by(これらの現象はすべて摩擦によって引き起こされる.)
   (4) as(それは、曲がったてことして働くと考えられる.)
   (5) must(運転手は彼のブレーキペダルを踏んだにちがいない.)
   (6) from(電気は電池または発電機から発生する.)
   (7) through(電気は回路の中を流れる.)
   (8) from, to(普通の自動車エンジンの rpm は 1000rpm から 5000rpm まで変えることができる.)
   (9) on(このシステムは十進法に基づく.)
   (10) from, to(技術は人によって異なるかもしれない.)
   (11) that(確かであるすべての整数を記録しろ.)
   (12) between, and(気体の圧力とその体積の間に関数関係がある.)
2. (1) Energy exists in a variety of forms.
   (2) Take a feather and a coin, and drop them from the same height.
   (3) In ancient times, people rubbed wooden sticks together to start fires.
   (4) We can calculate the power x exerted on the nail.
   (5) You all have seen a car suddenly stopped.
   (6) The reason we don' t notice its existence is that it doesn' t move.
   (7) When the electricity flows along an improperly closed path, it is called a "short circuit".
   (8) The exhaust gas from the gasoline engines contains contaminating substances.
   (9) The metric system is a system used for measurement in scientific and engineering work.
   (10) The measurement of a physical quantity is always accompanied with a degree of uncertainty.
   (11) There is no relation between the number of significant figures and the decimal point.
   (12) This can be expressed mathematically as $xy = a$.

Ⅱ 機械工学の周辺

1. (1) on(作業者はその正確な形と寸法に関する情報をもっていなければならない.)
   (2) in(正しい線が適切な場所で使われる.)
   (3) as, as, possible(この器具はできるだけ慎重に取り扱われなければならない.)

(4) into（それらは二つのタイプにざっと分類することができる.）
(5) of（限界ゲージは「go」側と「no go」側の二重のゲージである.）
(6) to, between, and（応力ひずみ線図は応力とひずみの関係を示す一つの方法である.）
(7) by, by（この強さは最初の断面積によって最大荷重を割ることによって得られる.）
(8) of, of（鋼は鉄と炭素の合金である，そして真鍮は銅と亜鉛でできている.）
(9) from（形状記憶効果はマルテンサイト相の変態により生じる.）
(10) with（鉄の棒で補強されたコンクリートは典型的例である.）
(11) as, as（炭素繊維の密度は普通 1.8g／cm^3 と同じぐらい小さい.）
(12) for（原油は，石油化学製品のための出発材料である.）

2. (1) A least two views are used.
(2) The figure of dimension is centrally written on the dimension line.
(3) The spindle of a micrometer is an accurately machined screw.
(4) Blocks are sold in a set.
(5) A limit gauge provides a rapid and accurate means of dimension measurement.
(6) It is very important to know the limit of endurance of materials to exterior forces or stress.
(7) Tensile strength is the strength necessary to pull a metal test piece into two pieces.
(8) An alloy is a metal mixture containing two or more metals.
(9) This strange behavior is called the shape-memory effect.
(10) Composite materials are novel materials composed of two or more materials.
(11) The principal use of high-performance carbon fiber is as a reinforcing component in composite materials.
(12) Crude petroleum is a mixture of various grades of hydrocarbons.

Ⅲ 機械工作

1. (1) in（機械はエネルギーを受け取り，そして役に立つ形で仕事を伝達しなければならない.）
(2) as（工作機械は金属を切断するための機械として定義される.）
(3) in（フライス盤はある意味では研削盤に似ている.）
(4) with（穴は穴をあける機械やボール盤のような適切な工具で作られなければならない.）
(5) of（研削盤は 2 種類ある.）
(6) to（せん断角はチップの厚さを決めるのに寄与する.）
(7) to（潤滑剤は部品の直接接触を妨げるのに用いられる液体または固体である.）
(8) by（抵抗溶接は電流によって発生する熱を使う.）
(9) to（鍛造は金属の塑性変形を意味する.）
(10) over（鋳造はほかの金属加工法に勝るいくつかの利点を提供する.）
(11) from（冷間圧延鋼は熱間圧延鋼からつくられる.）
(12) to（焼き入れは材料をより硬くすることを意味する.）

2. （1） A camera is not a machine for it never transforms energy through it.
（2） The lathe is the oldest and the most important machine tool.
（3） The milling cutter rotates while its workpiece is fixed on the table.
（4） The principles of its operation are nearly the same.
（5） The grinding machine is often used on metals too hard to machine otherwise.
（6） The material in front of the tool is continuously sheared along the shear plan.
（7） Lubricants mainly work by creating a considerably thick oil layer between two surfaces.
（8） The filter rod is usually made of the same material as the welded pieces.
（9） A metal has anisotropic structure when it is forged with very strong pressing force.
（10） Objects made by this method are called castings.
（11） The thick product of hot-rolling is called the bloom.
（12） Tempering means the re-heating of steel after the hardening operation.

Ⅳ 機械工学の現在

1. （1） for（CADはコンピュータ支援設計の頭字語である.）
（2） in（メカトロニクスという言葉は日本で最初に使われた.）
（3） to（センサは何かを感知する装置である.）
（4） to（フィードバック制御システムは望ましい値を得るためにコントローラを使う.）
（5） with（CNCはコンピュータ付きのNC工作機械です.）
（6） to（CAMはCADに密接に関連がある.）
（7） in（CIMはすべて情報がネットワークにつながるシステムである.）
（8） by（FMSはコンピュータの助けを借りてさまざまな製品を製造することができる.）
（9） by（燃料電池は化学反応で電気を発生させる.）
（10） like（太陽電池は火力発電のように燃料を必要としない.）
（11） but（ハイブリッド車はエンジンだけでなくモータももつ.）
（12） at（超電導は極低温で起こる.）

2. （1） CAD refers to the use of computers for designing products.
（2） The mechatronics is so-called Japanese English.
（3） A limit switch is a sort of sensor.
（4） The feedback control system is used to control the temperature of a refrigerator.
（5） CNC machine tool has become dominant in many manufacturing facilities.
（6） CAM is an acronym for computer-aided manufacturing.
（7） CIM is slightly different from CAD and CAM.
（8） FMS is a system by which various kinds of products can be manufactured on the same manufacturing line.
（9） The fuel cell produces electricity by a chemical reaction of hydrogen and oxygen.
（10） A solar battery is a device that generates electricity using the photoelectric effect of

the semiconductor.

(11) Since the hybrid car is fuel-efficient, it is called an eco car.

(12) A superconducting magnet is used in a linear motor.

V 管理技術

1. (1) in（部品はしばしば数百キロメートルの離れた工場でつくられる.）

(2) from（最も重要な要因はロットからサンプルを選ぶ方法だ.）

(3) by（検査によって拒絶された部品は捨てられなければならない.）

(4) on（寸法，温度，濃度のような測定された値はチャートに順番にプロットされる.）

(5) as（OR は最も良い解決策を決定するための科学的手法と技術の適用と定義される.）

(6) other（システム工学は二つの操作を含む：一つはモデリングで，もう一つは最適化である.）

(7) in, on（この概念は公害について前もって予測をするためにアメリカ合衆国で始まった.）

(8) by（PERT はコンピュータとほかの援助を用いて大規模のプロジェクトの時間予定を計画して，制御する方法である.）

(9) to（非破壊検査は非常に役に立つ試験方法であると考えられる.）

2. (1) The concept of interchangeability has established the foundation of mass production.

(2) Sampling inspection is the most powerful and effective method.

(3) Inspection involves the formal checking of the properties or dimensions of products.

(4) The theoretical background of the chart was proposed by Shewhart.

(5) OR is a decision-making process carried out by means of mathematical aids.

(6) They should be combined systematically to be treated by a computer.

(7) TA has a strong influence on various fields in Japan.

(8) PERT was developed for the Polaris missile project.

(9) Radiographic method is a method to detect flaws hidden inside the products.

課題英文要約

第Ⅰ章および第Ⅱ章は要約省略.

Ⅲ　機械工作

1　機械——機械は決まった運動をする剛体の集合で，有用な仕事をするものと定義されている．つまり，エネルギーを一方から受け取り，他方へ出すものである．この定義によると有能な仕事をするカメラは機械ではない．ところがコンピュータは可動部分もないし，エネルギーを変換することもしないが，現在では機械と認識されている．それは漠然とした情報を有用な形態に変換するからである．たとえば，コンピュータに情報を入力すれば解答が得られる．現在では機械とはエネルギーだけではなく情報をより望ましい形態に変えられるものと定義されている．

2　旋盤——旋盤は典型的な工作機械であり，刃物を固定し工作物を回転させる．私達はジェット機，船舶など多くの機械に取り囲まれており非常な恩恵を受けているが，それらはほとんど旋盤によって加工されている．旋盤は最古の最重要な工作機械である．工作物はチャックで保持され刃物に向って回転する．削られたくずはかみそりの刃のように危険である．さまざまな形状の工作物は工夫されたチャックに固定され，面削り，穴あけ，丸削り，ねじ切りなどの加工を受けることができる．

3　フライス盤——フライス盤もまた主要な工作機械で，その動作は研削盤に似ている．フライス盤は複数刃をもつフライスカッタという刃物を使用し，これは旋盤と違って，工作物が台の上に固定され刃物が回転する．工作物が送り込まれ，切りくずが生じる．大別して，(1) 横フライス盤，(2) テーブルが回る万能フライス盤，それに (3) 縦フライス盤にわけられる．フライス盤は，平面，曲面，不規則曲面，溝穴，溝，キー溝，カムなど多くの工作が可能である．ドリルの溝や歯車など複雑な曲面はらせん形の刃物で加工することができる．

4　ボール盤——機械工場では多くの穴あけが必要になる．あるものは極めて精密であるが，あるものは精密さを要しない．いずれの穴もボール盤のような単能機で為されるので，工場ではボール盤の知識と習熟を獲得することが重要である．刃物の多くはツイストドリルで，形状，材質，保持方法，ジグの有無にかかわらず原理はほぼ同じである．穴をたくさんあけるときは，ジグを用いるのが有利である．ジグは工作物の固定とドリルの案内を行うもので，それを用意するにはいくらかの先行投資が必要であるが，時間および労力の節約を十分に補って余りあるものである．

5　研削盤——研削は非常に精密でかつ美しい表面を作り，平面研削と円筒研削とにわけられる．緻密な組織の砥石車によって許容差の極度に小さい平面を作り，またほかの方法では切削できない硬い表面にも応用できる．平面研削はさらに横軸型と縦軸型とがあるが，前者が一般的である．工作物は万力または磁石で固定され，砥石車は水平軸に取りつけられ，1/1000mm の精度で上下し，切削の深さは砥石車の位置で決まる．テーブルの横方向の移動は切削速度といい，縦方向の移動は送りという．荒削りでは，速い切削速度と大き目の送りを用い，仕上げ削りでは遅い切削速度と細かい送りを用いる．

6　金属切削の機構——削りくずの出る切削加工のいくつかの様式を Fig. 18 に示し，その機構を Fig. 19 に示している．すくい角 α と逃げ角をもった刃物が t_1 の深さで工作物の表面を移動する．刃物の前方の材料は材料表面と ϕ の角度で切削面に沿って切削される．このせん断角とすくい角とが切りくずの厚さ t_2 を決める．t_1，と t_2 との比は切削比 r と呼ばれる．せん断角，すくい角，切削比との関係は $\tan\phi = r\cos\alpha/(1 - r\sin\alpha)$ である．切りくずの厚さを決定するものはせん断角であることがわかる．一般的にいってせん断角が大きいほど切りくずの変形が小さく，切削加工が円滑に進行する．

7　潤滑剤——潤滑剤は相対運動している部品の直接接触を防ぎ，摩擦や摩耗を減らす．また，潤滑剤は部品の冷却や防食のためにも重要である．潤滑剤には天然のものと合成のものが，また液体のものと固体のものがある．潤滑剤は部品の表面に薄い油の層をつくることによって，動いている部品の焼き付けを防ぐ．そのため，相対運動している部品の重さによって適切な粘度をもつ潤滑剤を選ばなければならない．

8　溶接——金属片の接合にはボルトやナットを使うほか，溶接が行われる．溶接は熱で金属片の辺を溶かし混ぜ合わせるのだが，多くは溶接金属と同じ材質の溶加棒を溶かして加える．溶接される金属を母材といい，加熱はガスの燃焼や電気による．金属の接合にはほかに鍛接または圧接がある．これらは接合する部分を溶ける寸前まで熱し，それに圧力を加える．抵抗溶接は小面積を通して流れる電流による発熱を利用したもので，溶接部分が線状でなく点状を示すのでスポット溶接と呼ばれる．

9　鍛造——鍛造は高温で圧力を加えて行う塑性加工である．単純なものは加熱金属を 2 枚の平板に挟んで平らにする加工で，据え込みといわれ，ハンマ，アンビル，スエジを用いる．大規模のものは油圧プレスを用いる．金属を大きな力で鍛造すると，木材のように繊維状の構造に変化する．このように鍛造された金属は丈夫になるので，スパナ，ニッパ，機関車の車輪，自動車のコンロッドなどに応用されている．鍛造のしやすさは，材料の強度やじん性などに左右され，アルミ合金，マグネシウム合金，銅合金，鋼の順に鍛造しやすい．

10　金属の鋳造——金属の鋳造とは，金属を溶かして型に流し込み，冷えて固まったところで型を取り除いて，製品を得る操作である．出来たものは鋳物であり，その表面は疎雑なので仕上げが必要である．鋳造の有利な点は大きな複雑な形状のものが大量生産できることである．鋳物は繊維構造をもっていないので鍛造品よりも強度で劣る．鋳型とは金属を流し込む型をいい，砂型または金属型で造られる．砂型は安いが一度限りであるのに対し，金属型は繰り返し使用でき，良好な表面が得られる．

11　冷間圧延鋼材——冷間圧延鋼材は，加熱時に圧延された熱間圧延鋼材からつくられる．熱間圧延鋼材は最終製品の寸法よりいくらか大きい．熱間圧延された製品の最厚のものはブルームと呼ばれ，それは熱間や冷間の圧延によって，ビレット，スラブ，板，薄板，箔へと薄く加工される．熱間圧延された鋼材は硫酸で表面の黒皮を除き，水で酸を洗い，石灰水に浸漬され乾燥される．その後磨きロールの間で非常な高圧をかけられる．冷間圧延鋼材は精密な厚さと美しい表面が得られるので，そのまま，あるいはさらに加工して使用される．

12　熱処理——熱処理は金属や合金を改善するために，溶けない程度に加熱あるいは冷却する操作をいい，焼き入れ，焼き戻し，焼きなまし，表面硬化がある．焼き入れは材料を硬化する操作で，炭素鋼を桜色になるまで熱し，空気，水，油などで急冷する．焼き戻しは焼き入れ鋼を再加熱し徐冷して望みの硬さとじん性を回復する．焼きなましは焼き入れの逆で，焼き入れ鋼を加熱後ゆっくりと冷却し，脆性を取り除く．表面硬化は低炭素鋼の表面に炭素を浸透させ表面にだけ硬さと耐磨耗性を与え，歯車などに実施される．

Ⅳ　機械工学の現在

1　CAD（コンピュータ支援設計）——CADは頭字語で，製品設計にコンピュータを使用することをいう．CADは，小はICの設計から大は大陸の地図の製作にまで幅広く応用され，さらに技術評価に必要なモデルまで製作する．従来の手作業に比べて，(1) 図面の変更，(2) データの部分的利用，(3) 複雑な図面の修正と再編，(4) 図面の拡大・縮小の点において優れている．CADの利用で製図作業の生産性は著しく向上した．以前はCADは大型コンピュータを用いたが，現在では16ビット位のパソコンで作動するので，近い将来には製図室内の図板や製図用具はパソコンとCADソフトで置き換えられるであろう．

2　メカトロニクス——メカトロニクスは和製英語であるが，現在は世界中で用いられている．機械に新しい機能を与えるために，機械工学と電子工学とを組み合わせた新しい工学である．機械の正確さ，迅速さ，巧妙さなどはかつてリンク装置やカムを使って達成された．現在ではセンサ，マイコン，アクチュエータなどの組合せが用いられる．典型的な例はロボット，マシニングセンタなどがある．システムエンジニアリングではスーパーコンピュータを使用した大規模なシステムであるのに対し，メカトロニクスは各機械に小さいマイコンを配し，ソフトウェア技術でその多機能ぶりを発揮している．

3　センサ——工作機械の運転は刃物の送りや切り込みの深さなどハンドルの目盛りで加減していた．機械の自動化につれて，人間の感覚の代わりに刃物や工作物の移動を自ら判断する装置，センサが使われるようになった．それは物理量の検出だけではなく，変換したり，ほかへ送り出したり，さらにときにはエラーを除去したり，補償したりする．電気機器で利用されるリミットスイッチは，センサの一種で他の物体との接触を感知することにより電気回路を開閉し，工作物の停止やスタートなどの動きを制御する．カメラの自動焦点もセンサの一種で，距離を測定し，焦点を自動的に決める．

4　フィードバック制御システム——開回路システムは制御するのに調節器を用いている．これに対して閉回路システムでは出力値と目標値との偏差を利用して制御している．出力を測定するセンサ部から，偏差を操作対象に加える調節器に至るまでの道筋をフィードバック回路という．このルートは信号を伝えるだけの道筋であるが，すべてのフィードバック制御系の心臓部である．自動車運転では運転手が自動車の位置を目的地と比較判断し，ハンドル操作を行う．つまりセンサ，調節器，アクチュエータの働きを兼ねている．家庭にある冷蔵庫などの電気機器は目標値をインプットするノブ，温度を測るサーモスタット，比較装置，調節器がフィードバック回路で結ばれている．

5　NC工作機械——NC工作機械はテープなどに記録されたデジタル数値によって自動的に制御される．すなわち工作物と刃物の相対位置，工具の移動速度，テーブルの速度などがメモリの中に蓄えられ，一連の命令信号として送り出され，それがサーボ機構を通して設計どおりに装置を作動させる．高性能のマイコンと高密度のメモリを持ったNC工作機械をCNC工作機械といい，その特徴は大量生産ではなくむしろ製品の多様性にある．最近マーケットの要望が「少品種大量生産」から「多品種少量生産」へと移ったことを考えると，CNC工作機械の将来がますます有望になるであろう．

6　CAM（コンピュータ支援製造）――CAMはCADと密接に関係している．従来の工場では職長が仕事の段取りをした．職長は仕事を受け取ると，機械，材料，工員などの手配をした．今日ではそのようなわずらわしい仕事はコンピュータに任せられるが，それをCAMという．さらに人事の管理まで行うようになった．労働者が突然休みを取ったとき，職長がキーボードを叩きさえすれば最も適切な処置が直ちに画面に現れる．CAMがもたらす利点は大量生産，時間の節約，製品の品質向上である．

7　CIM（コンピュータ統合生産システム）――CIMは生産活動最適化のため生産関連のすべての情報や機能をコンピュータでネットワークする．それは設計，分析，計画，購入，原価計算，在庫調査などをコンピュータで結ぶことを意味する．最近のCIMは市場調査，経営戦略，企業予測，意志決定までも含むようになった．CIM工場では部品や製品は停滞しない．それらは工程中か運搬の途中にある．従来の工場では加工時間は5％であるといわれるが，これからの工場は部品の貯蔵も製品の倉庫も要らない．そこではロボットが人間に代って部品などを運搬する．CIMは日本ではFAと呼ばれる．

8　FMS（フレキシブル生産システム）――FMSはコンピュータの力を借りて同一ライン上に多種の製品を流すシステムをいい，この柔軟性は今日では大量生産と同様に重要視されており，市場の要望にこたえて生き残るための絶対条件となっている．今までの硬直したラインと組織では社会のニーズにはこたえられない．理想的なFMSは流れる製品の種類が全部異なる．FMS工場ではCNC工作機械や自動組立機が使われ，ソフトの変更だけで異なる製品がつくられる．1970年にFMSが機械工場で採用され，現在では，衣服，食品，石油などの分野でも応用されている．

9　燃焼電池――水は電流を流すと水素と酸素に分解する．逆に，水素と酸素が化学反応すると電気と水を生じる．この原理を利用して電気を作り出す発電装置が燃料電池である．燃料電池は一つでは電圧が小さいので，普通たくさんの燃料電池が組み合わされてスタックという形で使用される．燃料電池にはアルカリ型燃料電池，溶融炭酸塩型燃料電池，りん酸型燃料電池，固体電解質型燃料電池，固体高分子型燃料電池があるが，固体高分子型燃料電池が家庭用燃料電池や燃料電池自動車に使用されている．燃料電池の長所は排気ガスとして水蒸気しか出さないので，環境にやさしいことである．

10　太陽電池――太陽電池は電池ではなく発電装置である．太陽電池は半導体による光電効果を利用して発電する．そのため，太陽電池はほかの発電装置とは異なり，燃料を必要としない，クリーンエネルギーである．そのうえ，太陽電池には可動部分がないため，信頼性も高いことが特徴である．太陽電池は光さえあれば発電できるが，太陽電池で発電される電流は光の量に大きく依存し一定ではないことに注意しなければならない．

11　ハイブリッド車――ハイブリッド車が普通の自動車と異なる点は，エンジンだけではなくモータをもつことである．ハイブリッド車はエンジンとモータの使い方で分類される．一つはモータが動力源でエンジンは発電にのみ使用されるタイプ，もう一つはモータとエンジンの両方が動力源として使用されるタイプである．このタイプには，エンジンが主な動力源であるものと，モータが主な動力源であるものがある．ハイブリッド車の長所は燃費がよいことである．ハイブリッド車はモータがあるだけではなく，回生ブレーキなど燃費をよくする工夫がされている．

12　超伝導——金属などに極低温で見られる現象で，特質特有の転移点で直流抵抗が突然ゼロになる．1911年にオンネスが水銀の電気抵抗の研究中に発見した．彼は4.2Kで抵抗がなくなることを発見した．1957年にJ.バーディーンらはこの現象を理論的に説明する電子配置を提唱した．1986年にこの現象がある酸化金属で35Kで起こることが発見され，その後90Kもの高温の転移点をもつ物質が見出された．転移点が高いと高価な液体ヘリウム使う必要がないので大変有利になる．超伝導による強大な磁力を利用して，大型加速器，リニアモータカー，SQUID，MRIなどが開発されてきた．

V　管理技術

1　互換性——自動車の何万という部品は数百キロも離れた工場でつくられることが多い．車体はある工場，エンジンは別の工場という具合である．したがってそれらは正確に同じ材質，同じ寸法でなければならない．正確につくられなければ寄せ集めた時に図面のとおり組み立てられない．完全な互換性が要求される．互換性の概念がない頃は，職人が部品をやすりですり合わせて組み立てていたのだが，現在では状況が変わった．検査を通ったすべての部品はむだなすり合せ作業は要らない．現在の繁栄を支えている大量生産は互換性のたまものであるといってもいい過ぎではない．

2　サンプリング——製品の検査をする際，簡単に検査できるものは全数検査をするが，品物が大量であったり，検査で品物が損傷するときは全数検査は不可能である．その場合抜取り検査が強力な手段となる．抜取り検査はロットから少数のサンプルを抜き取り，その検査の結果で元のロットを判断する．サンプルがよい結果を示せばロットを合格とする．抜取り検査を行う時に大切なことは，サンプルの抜取り方法であるが，それは抜取り者の作為をなくしたランダムサンプリングで行われる．そのために乱数表や乱数さいを用いる．

3　検査と試験——検査とは材料や製品などの性質を適当な時期に調べることをいう．完全な検査は設計の評価や設備の調整や使用時の耐久性なども含む．不正確な検査は粗悪な品物へとつながるので，検査器具も検査される．検査と試験とはほぼ同意で使われるが，検査はより広い概念で，仕様の解釈，製品の測定，サンプリングの方法などを含む．検査は検査部門の専用の部屋で行われる．空気調節や塵埃の調節も必要に応じて行われる．

4　管理図——管理図は製造工程が安定した状態にあるか，または緊急な措置が必要であるかを調べるもので，品質管理に欠くことができない．普通は中央線と2本の管理限界線でできている．寸法，濃度などの測定値を図の上に順次打点する．打点が著しい傾向を示せば工程が悪条件の下にあると判断される．管理図は統計学に基礎を置くもので，母集団は安定していることが必要である．母集団とは管理図作成のためサンプルを取る際の元のロットである．これは図の基礎を作った人にちなんでシューハート図といわれ，日本へはデミング博士によって紹介された.

5　オペレーションズリサーチ（OR）——ORはイギリス軍の作戦行動から生まれたことばである．第二次大戦中イギリスは苛烈なドイツ空軍の爆撃下にあり，その空襲を避けるための作戦行動がORの始まりだった．ORとは問題解決の科学的方法と技術である．または最適の選択を決定するための数学的手段ともいえる．それはよく「七つの未知数と，四つの方程式」ともいわれる．過程は，（1）問題の把握，（2）モデルの作成，（3）最適技法の選択，（4）最適解の発見，（5）実施である．大きな建物や学校などで，エレベータ，段階，トイレをいくつ，どこへ設置したらよいかを解決するにはORの助けを借りることが多い.

6　システム工学（SE）——システムとは目的達成のために組織内の関連し合う要素を有機的に結合したものを意味する．SEは工学の一部であって，まずシステムの要素を分解し，次にコンピュータなどの助力を得てこれを組み立てる技術をいう．それには二つの作業，つまりシステム内の各要素と実行の基準を明確にするモデルを作り，次に各要素を最善の遂行が得られるように調整することを含む．SEは日常の生活でも必要である．たとえば調理では栄養，経済，コックなどの要素が普通は主婦の処理に任されているが，大規模の調理ではシステムとしてコンピュータなどで調整されよう．SEの最も成功した例は日本の新幹線で，運転時刻，従業員，駅舎などの要素が見事にシステム化されている.

7　テクノロジーアセスメント（TA）——かつて技術の影響は常にプラスと判断され，公害などの影響が現れたときでも対策は後追いであった．最近は情況が変って悪影響は計画の段階で予想されるようになった．このマイナス面の評価をTAという．これは1967〜1968年にアメリカで公害を予想し，予防策を講ずるために始められた．日本でのTAプログラムは1970年にゴルフ場の農薬による水汚染などの予防対策から始まり，各方面に影響を与えた．第三世界ではTIA（技術影響評価）が問題になっており，これは先進国から持ち込まれるプロジェクトによるマイナスの影響を予想し防御策を講ずることをいう.

8　パート（PERT）——PERTとは，頭字語であり，大規模な事業の日程をコンピュータで予想し調整することをいう．これは1950年代後半にポラリスミサイルを成功させるためにアメリカで開発された方法で，その特徴は，（1）計画に含まれる無数の要素をできるだけ小さく解析する，（2）解析したものを矢線図を用いて順序づけする，（3）相互の関連を調べる，（4）図上で障害となる要素やパスを確認する，（5）障害は実行する前に調整される．PERTの利点は計画段階で障害を発見し処理することにある．これまでは要素間の依存関係は必ずしも明瞭ではなかった．PERTを用いれば網状構造や経路を解析することで問題点が明らかになり取り除くことができる.

9　非破壊検査——製品を破壊することなく欠陥や性質を調べる有用な方法である．医者が用いるX線や超音波が工業でも応用される．目視探傷法は肉眼による観察であり，特殊な場合にはレーザ光や光ファイバを利用する．放射線透過法は内部の欠陥を探すのにX線，ガンマ線などを使う．超音波法は内部の欠陥や不連続部分から反射される超音波を利用する．侵透探傷法は染料などを欠陥部分に浸み込ませそれを現像して欠陥を探し出す.

索　　引

ア行

カ行

ナ行

ハ行

ワ行

英字

〈著者略歴〉

青柳忠克（あおやぎ　ただかつ）

昭和 17 年　横浜高等工業学校応用化学科卒業
昭和 48 年　東京都立鮫洲工業高等学校校長
昭和 58 年　東京工業大学講師
元著述業，翻訳業

齋藤哲治（さいとう　てつじ）

昭和 59 年　京都大学工学部金属加工学科卒業
昭和 61 年　京都大学大学院工学研究科修士課程修了
平成 9 年　千葉工業大学工学部助教授
平成 15 年　千葉工業大学工学部教授
博士（工学）

塚原隆裕（つかはら　たかひろ）

平成 15 年　東京理科大学理工学部機械工学科卒業
平成 19 年　東京理科大学大学院理工学研究科博士課程修了
平成 20 年　東京理科大学理工学部機械工学科助教
平成 25 年　東京理科大学理工学部機械工学科講師
博士（工学）

やさしい機械英語（改訂２版）

1994 年 1 月 25 日　第 1 版第 1 刷発行
2016 年 11 月 25 日　改訂 2 版第 1 刷発行
2025 年 4 月 10 日　改訂 2 版第 6 刷発行

著　者　青柳忠克
　　　　齋藤哲治
　　　　塚原隆裕
発行者　髙田光明
発行所　株式会社 オーム社
　　　　郵便番号　101-8460
　　　　東京都千代田区神田錦町 3-1
　　　　電話　03(3233)0641(代表)
　　　　URL https://www.ohmsha.co.jp/

印刷　美研プリンティング　　製本　協栄製本
ISBN978-4-274-21974-0　Printed in Japan